ABOUT TIME
CELEBRATING MEN'S WATCHES

IVAR HAUGE LINE

Schiffer Publishing Ltd®

4880 Lower Valley Road • Atglen, PA 19310

Other Schiffer Books on Related Subjects

Rolex Wristwatches: An Unauthorized History, 3rd Edition by James M. Dowling & Jeffrey P. Hess, ISBN 978-0-7643-2437-6, $125.00

Sports Watches: Aviator Watches, Diving Watches, Chronographs by Martin Häussermann, ISBN 978-0-7643-4599-9, $39.99

The Fascination of Time: Marks, Manufacturers, & Complications of Classic Wristwatches by Harry Niemann, ISBN 978-0-7643-4685-9, $34.99

ISBN: 978-0-7643-4905-8
Printed in China

Published by Schiffer Publishing, Ltd.
4880 Lower Valley Road
Atglen, PA 19310
Phone: (610) 593-1777; Fax: (610) 593-2002
E-mail: Info@schifferbooks.com

For our complete selection of fine books on this and related subjects, please visit our website at www.schifferbooks.com.
You may also write for a free catalog.

This book may be purchased from the publisher.
Please try your bookstore first.

We are always looking for people to write books on new and related subjects. If you have an idea for a book, please contact us at proposals@schifferbooks.com.
Schiffer Publishing's titles are available at special discounts for bulk purchases for sales promotions or premiums. Special editions, including personalized covers, corporate imprints, and excerpts can be created in large quantities for special needs. For more information, contact the publisher.

Design: Aldente Advertising
Cover photo: Fredrik Ringe
Font: Baskerville Regular and Alternate Gothic EF-No Thr Regular

ABOUT THE AUTHOR
When he was 15 years old, Ivar Hauge Line discovered one of his greatest passions: watches. Over the years, as a collector, he has become part of a large network of manufacturers, watchmakers, and other enthusiasts. Ivar visits watchmakers and shops across the globe and attends the world's largest watch show in Switzerland each year, Baselworld. Today he is part-owner of a watch distribution company, Watchline, and the happy owner of more than 250 wristwatches.

CONTENTS

6	WHAT TIME IS IT, REALLY?	
10	THE SAVIOR OF THE SWISS WATCH INDUSTRY	6
16	10 WATCHMAKERS. 3 YEARS. DESIGNED IN 1783. FINISHED IN 2007.	10
18	THE NAME BEHIND THE WATCH	16
20	ULTIMATE WATCH COLLECTION FOR MEN	18
42	BLANCPAIN FIFTY FATHOMS	20
51	ALBERT	42
52	THE WORLD'S FIRST SMART WATCH	51
54	WATCHMAKING TRADITIONS IN FLORENCE	52
58	EL PRIMERO — THE WORLD'S MOST STUNNING MOVEMENT	54
66	ZENITH AND FELIX — A SUPERSONIC COUPLE	58
74	SPACEMASTER Z-33	66
76	CARS AND WATCHES	74
110	BRUVIK — A PIECE OF NORWAY	76
114	IT TAKES COURAGE AND PERSEVERANCE	110
116	OMEGA ON THE MOON	114
130	FRANCK MULLER	116
132	TIME IN WORDS	130
134	PILOT WATCHES	132
142	PORSCHE DESIGN INDICATOR	134
143	HUBLOT ONE MILLION $ BLACK CAVIAR	142
144	BREITLING EMERGENCY	143
146	TAILOR-MAKE YOUR OWN WATCH	144
154	PAUL NEWMAN AND HIS DAYTONA	146
162	ROLEX MOVES FROM ENGLAND TO SWITZERLAND	154
163	CONCORD C1 TOURBILLON GRAVITY	162
164	CERTIFICATION	163
165	EIGHT PAST TEN	164
166	MINUTE REPEATER AND PATEK PHILIPPE SEAL	165
168	CELEBRITIES AND THEIR WATCHES	166
170	WATCHES' ROLES IN MOVIES	168
178	PANERAI L'ASTRONOMO LUMINOR TOURBILLON 1950 EQUATION OF TIME	170
180	MONDAINE & APPLE	178
182	PLOPROF — OMEGA'S FIRST DIVING WATCH	180
192	RED ADAIR	182
194	SPACE INVADERS	192
198	BREITLING NAVITIMER	194
200	TUDOR — IN THE SHADOW OF ROLEX	198
206	PATEK PHILIPPE	200
210	PATEK PHILIPPE 324 S QA AND THE GENEVA SEAL	206
212	TOURBILLON AND GREUBEL FORSEY	210
216	WATCH ADS	212
218	HM4	216
220	WATCHES SOLD AT AUCTIONS	218
224	JAEGER LECOULTRE AND U.S. NAVY SEALS	220
226	HYT H1	224
228	THE WATCH THAT UNLOCKS YOUR CAR	226
229	CAR BRANDS AND WATCHES THAT MATCH	228
230	BASELWORLD	229
232	SMALL SCULPTURES	230
233	COMPLICATIONS AND EXPLANATIONS	232
		233

Today a watch is more
of a status symbol than
a time-telling gadget.
The watch on your wrist
tells me a lot about
who you are...

WHAT TIME IS IT, REALLY?

FOR SOME OF US, WATCHES ARE BOTH PASSION AND SCIENCE. PERSONALLY, I HAVE BEEN FASCINATED BY WRISTWATCHES SINCE I WAS A TEENAGER. WHEN I MEET NEW PEOPLE, I ALWAYS NOTICE WHAT KIND OF WATCH THEY WEAR AND FIND I ENJOY OBSERVING WHAT THAT TELLS ME ABOUT HIM OR HER.

When traveling, I always find time to stop by a few watchmakers or shops. Some people might bring 8 pairs of shoes or 4 suits on their journeys, while I always carry 4 to 5 watches in my luggage.

Everyone has a relationship to watches. Who doesn't remember their first watch? They are passed through generations, worn as jewelry or to show off, and used in ads; "Every Rolex tells a story." Watches are a natural gift to someone you love or a gift to yourself. From father to son, or from son to father. Watches can be a reward for long and faithful service, serve as an honorary gift, or mark a sports achievement.

Different watches usually follow specific professions. The architect doesn't buy the same watch as the broker. A particular watch can be used as a means to show off how successful you feel you are, but also give an image of who you want to be. After we heard James Bond say "Shaken, not stirred" on the big screen, wearing an Omega Seamaster, the sales of that watch increased substantially. Others would like to appear more subtle, preferring to wear an expensive, less-known watch, to send out signals only a watch expert will catch.

I am among enthusiasts at a watch launch in Stavanger, Norway.

If you ask me, everyone should own more than one watch (I prefer to have a lot, to be honest, but I'll come back to that). Watches are fashion; trends come and go, in the watch industry like everywhere else. You don't wear the same watch when dressed up as you do every day. Your favorite watch fits most occasions. Men often have a sports watch that can handle a deep water dive, or at least a dip in the bathtub.

A watch isn't necessarily a device we use for practical reasons, like telling the time. If we are concerned about getting to work on time, chances are we check our PC, mobile phone, or the clock in the car. The wristwatch expresses who you are. It is the man's jewelry.

For the same reason, some of us buy watches with functions we never use. Who really needs a helium valve on their watch? (That's a valve that assures the helium is let out of your watch in exactly the right proportions while you're in your pressure chamber.) GPS, perpetual calendar, or GMT can be useful features. The flyback function may not apply to us all (unless you're into acrobatic aviation), but is seriously cool.

The quality of most watches today is good. One manufacturer delivers its movements to a number of different watchmakers, so the prices tend to

Watches are fun. That's a fact. For some of us they are a passion, a science and a really big interest. I love wristwatches and timepieces and have been excited about them since I was a teenager.

reflect image more than the actual quality. But not always. Sometimes there is a good reason why a watch is pricey. It may contain a complicated movement developed especially for that watch. Some watches are limited editions and some are even put together by hand.

There is something extra special about Swiss mechanical watches for the über-interested. They are not more accurate than traditional quartz watches. They are not cheaper. Still, "Swiss made" is an important point. If you study the technique and precision of a mechanical watch, painstakingly put together with hundreds of parts, it isn't hard to understand why: springs, pendulums, tiny screws, and stones. Complicated functions like perpetual calendars, moon phases, tourbillons, and mechanical alarms. Like music to any watch fanatic's ears.

Storytelling is a great part of the passion for watches. Why does James Bond flash an Omega? What kind of watch was Neil Armstrong wearing when walking on the moon? How long and how many watchmakers did it take to make a watch for the French queen Marie Antoinette (who was actually convicted for treason and executed before she got a chance to see the magnificent watch)? The ultimate collectors' watches, my own favorites, the celebrities' choices, and the watches' roles in famous movies... In this book, I'll tell you stories about watches you are familiar with—and some you might not have heard about.

If you are interested in soccer and your team is Manchester United, chances are you dream about talking tactics with Sir Alex Ferguson. To me, the dream was to meet Nicolas Hayek Sr. (second from left) and get the chance to lunch with him. Together with Knut Lervik, at the time president of the Norwegian Watchmaker Society, I met Hayek and his assistant, Francois Thiébaud. I remember Hayek was wearing four watches on each arm (!) and smoked big fat Cuban cigars.

THE SAVIOR OF THE SWISS WATCH INDUSTRY

NICOLAS G. HAYEK IS OFTEN REFERRED TO AS "THE MASTER OF TIME," "MR. SWATCH," AND "THE SAVIOR" OF THE SWISS WATCH INDUSTRY.

Hayek was also CEO of Breguet, the founder of Swatch Group. He said "a good entrepreneur is like an artist who constantly needs to create something new." Based on this philosophy, Swatch Group, the company Hayek created from the ashes of the Swiss watch industry in 1980, developed into one of Switzerland's greatest success stories.

Hayek was born in 1928 in Lebanon. He was the son of an American dentist who worked at the American University in Beirut and a Lebanese mother. Hayek graduated from University of Lyon in France and moved to Switzerland after completing his studies. In 1957 he married and had two children. When he lived in Zurich, he pledged his family's furniture towards a 4,000 franc bank loan. This was the first and last time he asked a bank for a loan (or anyone else for that matter). Hayek Engineering was founded in 1963 and developed into one of the most prestigious and successful Swiss management consulting companies.

Switzerland spent a vast amount of money on research and development of the quartz watch in the 1960s and '70s, before the Swiss watch industry plummeted into crisis and the Japanese watch industry took over the market.

The Japanese launched the quartz watch with great success and the sale of Swiss watches fell 25% in 1982 alone. At the end of the seventies, SSIH (Société Suisse pour l'Industrie Horlogère), the company responsible for Omega, Blancpain, F. Piguet, and Tissot, was declared insolvent. This was due to recession, Asian quartz, and poor management. At the same time, the U.S.A.'s cheap brand Timex claimed parts of the market. SSIH was once the third largest watch manufacturer in the world and the largest one in Switzerland.

Hayek and Breguet introduced the remodelling of the famous Marie Antoinette watch from 1783 during Baselworld. Read about this remarkable watch on page 17.

Another Swiss company, ASUAG (including ETA, Oris, Longines, and Rado), which had been the world's largest manufacturer of pocket watches, also failed in 1982, and was taken over by creditor banks. Under Hayek's recommendations, the two companies were reorganized and merged to form the ASUAG-SSIH Holding Company in 1983. Most people within the Swiss watch industry panicked and were ready to give up.

Several Swiss banks hired Hayek as a consultant and financial advisor in order to save the Swiss watch industry during a time when Japanese investors were hunting for renowned Swiss watch brands at a low price. One Japanese company offered 400 million Swiss francs for Omega. Mr. Hayek dissuaded a sale, even after the bid was raised to include an additional 5 Swiss francs for each sold Omega.

This watch industry would have looked very different if they had sold at the time. Omega was a famous and strong brand name in the entire industry and also in the market. Blancpain and Frederic Piguet had a different destiny and were sold to Jean-Claude Biver (with help from Jazques Piguet, the grandchild of the well known Louis-Elysèe Piguet) at a humiliating price of 20,000 Swiss francs. The Swatch Group bought both watch brands back in 1992 at one thousand times its sales price in 1981. Oris was also liquidated and taken over by two longterm Oris leaders and is still independent.

Hayek studied the Swiss watch industry and was certain the only way to save it would be to introduce a product like Swatch. For Hayek, this meant full automation, aggressive marketing, and a solid product. These were the key elements for Swatch's huge success. In contrast to conventional watches, a Swatch had 51 components,

Swatch Group has opened several
Swatch brand stores around the world.

about 55% less than ordinary similar watches. With a streamlined production, a Swatch ended up costing 80% less than the production of other Swiss quartz models.

It's no coincidence that the Swiss flag is incorporated in the Swatch logo. Swatch was a very emotional product, not just a commodity. This was Mr. Hayek's and the country's new pride. The success paved the way for other success stories of mechanical watches only a few years later.

In 1985, Hayek and a group of investors bought (for 300 million Swiss francs) the majority of the shares in Swatch Group, took it off the stock exchange, and privatized it. Hayek was elected CEO. The group was renamed SMH in 1986, and finally Swatch Group in 1998.

The Swiss watch industry was recovering. Classic brands like Omega, Hamilton, Rado, and Longines, amongst others, were given a new chance to shine.

Swatch Group also bought the majority of the shares of the brand Blancpain from Biver in 1992. Biver kept running Blancpain and helped administrate Omega until 2001 when Hayek's grandchild Marc A. Hayek took over as CEO.

In 1999 Swatch Group bought the brands Breguet and Lemania from Investcorp.

In 2002 Swatch Group announced they would phase out the sale of ebauches (partly mounted mechanical movements) within 2006. Most manufacturers do not make their own. This responsibility is left with ETA, F. Piguet, and Lemania. Hundreds of watch brands with no connection to Swatch Group all over the globe have their movements made by these producers.

The industry feared dramatic consequences when deliveries of ebauche movements were drastically reduced. The Swiss Competition Commission was involved and managed to come up with a compromise.

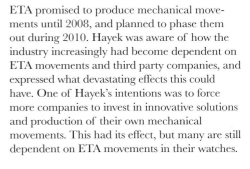

Jeremy Scott designed these three models: Winged Swatch, Lightning Flash, and Swatch Opulence.

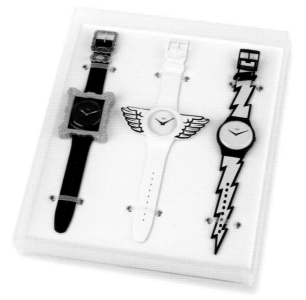

ETA promised to produce mechanical movements until 2008, and planned to phase them out during 2010. Hayek was aware of how the industry increasingly had become dependent on ETA movements and third party companies, and expressed what devastating effects this could have. One of Hayek's intentions was to force more companies to invest in innovative solutions and production of their own mechanical movements. This had its effect, but many are still dependent on ETA movements in their watches.

Hayek Sr. stepped down from his position as CEO in 2002 and let his son Nick Hayek Jr. take over. Nicolas Hayek received a number of prizes and awards before he died suddenly on June 28, 2010, in Switzerland, at the age of 82.

I was lucky enough to meet Hayek Sr. with Francois Thiébaud in 1992 over lunch in his office. I was fascinated by Mr. Hayek wearing four watches on each arm and remember him as a warm and welcoming person.

HAYEK & SWATCH

1996
Hayek carries the Olympic flame
through the streets of Atlanta.

1998
Hayek and Swatch Group launch
the smart car, together with
Mercedes Benz.

1979
Hayek enters the watch industry
when he is asked to save the
companies SSHI and ASUAG.

1985
Hayek invests in SMH and secures
the majority of the shares.

1982
After the merging of the two
companies, Hayek creates SMH
and launches Swatch.

swatch+

1990
Swatch becomes a collector's
item and "Jellyfish" is sold at
auction for $17,000.

1998
Swatch Group launches Internet
Time and their series with a
new time zone.

1984
Swatch engages famous
artists like Keith Haring.

1992
Swatch reaches record sales
of 100 million Swatch watches.

1998
SMH becomes Swatch Group.

1983
Swatch, as one of the first watch
manufacturers, opens its own
brand store.

1995
James Bond changes watch brand,
choosing SMH's Omega.

007

1986
Hayek becomes director and
chairman of the company.

1999
Hayek buys historic Breguet and continues its development of Double Tourbillon.

2007
Swatch introduces Hamilton Ventura or the "Elvis watch" for its 50th anniversary.

2009
Omega launches a limited edition Omega Speedmaster for the 40-year anniversary of Apollo 11's moon landing.

2010
Swatch continues to produce watches with well-known designers like Cary Card, Cassette Playa, and Manish Arora.

2006
Swatch launches a new series named Jelly in Jelly, to celebrate 333 million watches sold.

2008
George Clooney is elected as a board member in Hayek's environmentally conscious company Belenos Clean Power.

2004
Swiss court asks Hayek to keep producing ETA movements until 2010.

2010
Nicholas Hayek Sr., one of Switzerland's most famous personalities, dies at age 82.

2008
Hayek and Breguet launch a rebuild of the famous Marie Antoinette watch from 1783 during Baselworld.

2002
Hayek announces ETA will stop producing movements for other watch manufacturers during 2006.

2000
Swatch Group buys historic German watch manufacturer Glashütte Original.

10 WATCHMAKERS.
3 YEARS.
DESIGNED IN 1783.
FINISHED IN 2007.

MARIE ANTOINETTE, QUEEN OF FRANCE, STRIVED TO ALWAYS BE THE FIRST WITH THE LATEST. IN MANY WAYS A SPOILED CHILD, SHE GREW UP IN A SHELTERED AND PROTECTED WORLD IN VERSAILLES'S BEAUTIFUL SURROUNDINGS WITHOUT KNOWING OR UNDERSTANDING ORDINARY FRENCH LIVES.

When word reached her that there was a shortage of bread for the people she allegedly uttered, "Give them cake." A watch was commissioned by one of her admirers from watchmaker Breguet in 1783. Nothing was ever good enough for the beautiful queen, and all parts possible were to be made in gold. The watch was to have as many complications as possible and price was no issue. Watchmaker Abraham-Louis Breguet was commissioned to design the watch.

The Queen never received the extraordinary watch, nor did she get to see it. The French revolution started in 1789 and she was executed in 1793, at age 38. Breguet never had the pleasure of seeing the finished result either. It took 44 years to complete the watch and Breguet passed away a few years before it was finished.

This special golden watch ended up in a museum in Israel and was stolen in 1983. Fortunately, all drawings and design notes had been stored in Breguet's archives. When Nicolas G. Hayek took over the company and it was incorporated in Swatch Group, he was commissioned to make a copy of the watch.

Ten watchmakers spent three years recreating the masterpiece from the drawings and notes in archives from Breguet and from the Musée des Arts et Métiers in Paris. When Breguet opened their store in Place Vendôme in Paris, the watch, still not entirely finished, was exhibited behind thick bulletproof glass. It was presented to the public and press during Baselworld. Mr. Hayek opened a beautiful wooden box, made from an oak tree outside the Royal Chateau in Versailles, and finally the world was allowed to see but not touch.

The watch is encased in gold. It is a "Perpétuelle," both mechanical and self-winding, and consists of 823 different parts. The watch has a power reserve of 48 hours, a perpetual calendar, center seconds, thermometer, and jumping hours. An incredible sight!

Marie Antoinette,
Queen of France.

THE NAME BEHIND THE WATCH

It can be difficult to find a name for a new watch. One can use names describing the functions of the watch: Tourbillon and Perpetual Calendar from Patek Philippe, Porsche Worldtimer, Emergency from Breitling, Golden Bridge from Corum, Rolex Oyster Perpetual, Seiko Kinteic, Sea God GMT, or Jeager LeCoultre Reverso. These names require some

knowledge, but to a watch enthusiast they tell you a great deal about the timepiece.

Many watches are named after places in the world. This could of course be a little risky, especially if the chosen place obtains a bad reputation, since the idea is to give the watch positive associations. Examples include Portofino from

IWC, Cape Cod from Hermès, Hamptons from Baume & Mercier, or Jacques Lemans' Roma. Some models are named after famous races or race tracks: Le Mans, Mille Miglia, GP 1966, Sebring, Monza, and Silverstone.

Several watch models have become so well known that the name itself obtains a certain cult status and is referred to without the brand name. Here is a little test for you. What brand does the watch model belong to? Reverso, Monaco, Fifty Fathoms, Portuguese, Navitimer, Submariner, Luminor, Royal Oak, Big Bang, Speedmaster, Tank, Sea Dweller, Daytona, TwentyFour, and El Primero.

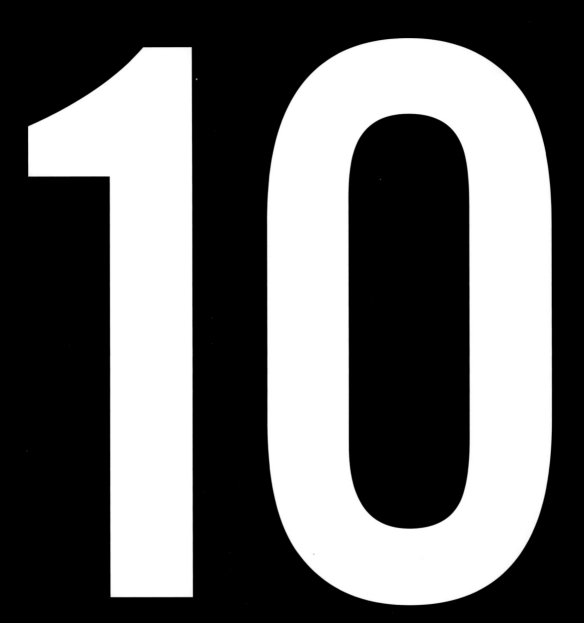

ULTIMATE
WATCH
COLLECTION
FOR MEN

Watches are all about fascination and feelings. It is seldom a rational choice. Okay, you may need a special function, but in the end it all comes down to what you like and what suits you. I relish technical complications and design. I'm not talking about price, it's more about what I like. I started collecting wristwatches more than twenty years ago, but still am missing quite a few to complete my wish list. Collecting is about wishing and searching for special pieces. I don't think I will ever reach the point where I tell myself I have all I want. I have consciously not put price tags on these watches, as the prices will vary depending on what edition you choose.

But these are relatively expensive pieces and some will stay on my wish list forever. There are a lot of wonderful watches I would like to have included in the list... Tudor Heritage Chronograph, Breitling Emergency, Hublot Big Bang Aero, Patek Philippe Nautilus, Porsche Design Indicator, Fortis Flieger Chronograph, and Omega Seamaster Planet Ocean Chronograph.

This is my personal top ten ultimate watch collection. They are listed randomly, but if you are building up a collection, I suggest you start with the legendary "moon watch"—Omega Speedmaster Professional.

PATEK PHILIPPE 5270 PERPETUAL CALENDAR CHRONOGRAPH

Charlie Sheen

This watch and brand is for connoisseurs—for those who can afford an expensive but discreet watch. Patek Philippe fits like a tailor-made suit. Impeccably dressed at all times. This model is, despite the price, almost impossible to get hold of: Patek Philippe model 5270 including perpetual calendar and chronograph (5270G). In white gold, black croco strap, with a movement measuring 41 mm wide, this is not a spectacular watch. It is simply classic and elegant. I am pleased this model measures more than 40 mm and not 38 like many others. The watch is easy to read with date, day, and month. In addition, it has the traditional chronograph functions and the option to show the time in seconds, minutes and hours. Patek Philippe is a typical collector's item. The price will most likely increase, and because of its timeless design, it will still be a great watch in twenty years' time. This is the watch you don't own, but keep for the next generation, as stated in their ads.

ZENITH EL PRIMERO STRIKING 10TH

Zenith ambassador Johan Ernst Nilson conducted an expedition from the North Pole to the South Pole.

While ordinary automatic mechanical watches have eight beats per minute, Zenith El Primero Striking 10th has ten—which makes it one of the world's best and most beautiful movements. I am talking about the motor of the watch. The El Primero 4052B automatic model comes in a limited edition of 1,969, with steel or leather strap. The movement measures 42 mm, equipped with a power reserve of 50 hours; it also has a chronograph that allows you to time down to one-tenth of a second. Then there's the characteristic vivid red second hand with the Zenith star on one end, and a traditional three-color sub-dials design of silver-gray, black, and dark blue. This brand is bursting with history. Gandhi and Amundsen both wore Zeniths and the company almost disappeared when the American Zenith company decided to end their watch manufacturing. One of the employees was smart enough to collect drawings, designs, and tools, and managed to start producing a few years later. Too bad this brand is not wider known, because it deserves it. Personally, I would choose the model with the steel strap.

Zenith El Primero Striking 10th Chronograph.
Limited edition of 1,969.

Audemars Piguet Royal Oak Offshore Chronograph Tourbillon.
Automatic, 34 rubies, carbon and ceramic with
a diameter of 44 mm. Rubber strap.

AUDEMARS PIGUET ROYAL OAK

Designed by the legendary Gérald Genta in 1972. Within four decades, the Royal Oak has become the backbone of Audemars Piguet's collection. The unmistakable iconic octagonal bezel with inspiration from the eight British HMS *Royal Oak* ships... This is a watch I like a lot and my favorite (also on my wish list) is a Royal Oak Offshore Chronograph. A cool and chunky men's watch, with the same characteristic octagonal shape and eight screws on the bezel with a 44 mm large case. Big push pieces, and a steel strap or a heavy rubber strap. Ceramic bezel and sapphire glass. Caliber 3126/3840. Over time, AP and Royal Oak have grown, become more daring and more complicated, and still remain true to their roots as a complete men's watch. If I could choose whichever watch I wanted, I would pick this watch in rose gold with a black leather strap.

Customized Audemars Piguet Royal Oak Offshore Arancio by Titanblack

IWC PORTUGUESE

IWC Portuguese is a unanimous tribute to the pioneers of Portuguese seafaring: Vasco da Gama, Bartolomeu Dias, and Ferdinand Magellan, who still today are honored for their outstanding sailing skills.

IWC also has a long history and makes fantastic watches. I have two favorites: Big Pilot and Portuguese. If I had to choose one over the other, the choice is Portuguese. "Heroes of the sea, noble people…" is the opening line of the Portuguese national anthem. It expresses a unanimous tribute to the pioneers of Portuguese seafaring, Vasco da Gama, Bartolomeu Dias, and Ferdinand Magellan, who are still today honored for their outstanding sailing skills, precise nautical charts, and use of instruments. This watch is IWC's tribute to them. It combines traditions and nautical instruments with modern design and futuristic mechanics. The watch comes in several different designs: the complication-rich Grande Complication, Minute Repeater, Tourbillon, Sidèrale Scafusia, Yacht Club, and the one I would choose: Perpetual Calendar. Anthracite face and brown crocodile strap. It is a limited edition of 250, and has a 7-day power reserve. The highlights include the perpetual calendar, showing date, day, month, and year in four digits, and a moon phase. The case is made of white gold and has a diameter of 44.2 mm.

IWC Portuguese Perpetual Calendar Ref. 5025.
Limited edition in platinum, 7-day power reserve,
44.2 mm diameter, and a perpetual calendar.

Jaeger LeCoultre Reverso Grande GMT in stainless steel.
24 hour indicator, 8-day power reserve, and GMT time
difference, with brown alligator strap.

JAEGER LECOULTRE REVERSO

Clive Owen with his Jaeger LeCoultre
in connection with Reverso's
80th anniversary.

Just as unique today as when it was first introduced. The Jaeger LeCoultre reversible watch was of course named Reverso. It means that you can turn the case and view another watch with a different time zone or another function. This is the watch that Don Draper, very timely, is wearing in *Mad Men*.

Whether it is to your liking or not, it is a timeless, classic piece and goes well with a suit. It was first made in 1931 for polo players in India. The watch can be reversed to protect the dial. Reverso is still almost identical to the original watch, but today the reverse is often used for secondary functions like moon phase or a different time zone. The model is made in stainless steel with a brown alligator leather strap. It is hand-winding and has an 8-day power reserve. The dial features hours, minutes, and seconds, and it has a Jaeger LeCoultre 822-caliber movement.

Clive Owen with his Jaeger LeCoultre in connection with Reverso's 80th anniversary.

ROLEX SUBMARINER

Submariner is still one of the most classic and well-known models from Rolex.

Right after the Omega SeaMaster, the Submariner is the best selling watch in Scandinavia and *the* best selling in Europe. There is no doubt this is Rolex's greatest success. Personally, I am not a Rolex fan, but vintage Rolex models like Daytona, Sea Dweller and Submariner I quite like. You simply cannot ignore this iconic brand. Rolex Submariner was launched in 1953, and was the first watch that was guaranteed waterproof down to 100 meters; it rose to stardom after Cousteau's deep-sea diving use. In the sixties, Rolex introduced another innovation, the Submariner model: featuring the world's first helium valve. This is the watch the first divers in the North Sea used during the age of pioneering the Norwegian Continental Shelf. The French diving company Comex also equipped their divers with this watch and had their own series made with the Comex logo printed on the dial. Submariner has remained more or less the same since it was introduced. This is one of the great classics of diving watches. And remember: stainless steel with black bezel only.

Rolex Submariner Date in stainless steel with black bezel. 40 mm diameter, automatic, with stainless steel strap.

Omega Speedmaster Professional Moonwatch Apollo 40th Anniversary Limited Edition. With "02:56 GMT," the exact time Armstrong set his foot on the moon on July 21, 1969.

OMEGA SPEEDMASTER

The classic moon watch has got to be on the list. This is the first watch you should buy when starting a collection. Introduced as a sport chronograph in 1957, the Speedmaster was chosen by NASA as their official chronograph, certified to travel in space, and was the first watch to be worn on the moon in 1969. Even if the design has changed somewhat, Speedmaster remains NASA's officially chosen watch, and is still the only watch that has been certified for the moon. The original piece was mechanical and Omega caliber 321. In 1968 it was replaced by caliber 861, an upgraded and improved version. The watch is still the same, with its signature stainless steel strap and black bezel/tachymeter scale Chronograph version with hours, minutes, and seconds. Slim, white illuminated hands. This watch has become a collector's item and has been launched in different series and versions in limited editions. Older models are selling at high prices in auctions. The case is 42 mm and the watch is water resistant down to 50 meters.

Buzz Aldrin with an Omega Speedmaster Professional on his wrist.

PANERAI RADIOMIR

Orlando Bloom with Panerai Black Seal.

I love Italy and have a special relationship with Florence. This is design and history from Italy with Swiss precision and mechanics. Perhaps it is not such a surprise that I like Panerai, which started its life in Florence. The first Panerai Radiomir was conceived in 1935, when the Italian Marine contacted Panerai and asked them to manufacture watches for their special forces. Many watches for men are perfectly fine for a woman to carry as well, but this is a watch for men only. This beautiful watch is both masculine and elegant at the same time. Panerai has kept its style and conic cushion shape structure. It was originally designed for rough environments, and is now considered one of the world's absolute classics. But even the most timeless pieces could do with the occasional upgrade. My choice is Panerai Radiomir 10 days GMT. Known for its grand sizes, this is no exception, measuring 47 mm. A beautiful brown dial with sand-colored numerals and indicators. Indicator for 10-day power reserve, date, and two time zones—GMT. I like this watch best with a brown, worn patina alligator strap.

Panerai Radiomir 10-day power reserve, date, and two time zones. 47 mm diameter with brown leather strap.

Tag Heuer Monaco 24 Limited Edition 40.5 mm diameter. Black titanium case in stainless steel with Tag Heuer Caliber 36 movement.

TAG HEUER MONACO

I have a thing for classic and historical watches. Tag Heuer Monaco is one of these. This is the watch Steve McQueen wore on his wrist in the legendary car movie *Le Mans*. He made this Tag Heuer's decidedly best-known timepiece. This model was also the first square chronograph the world had seen when it was launched in 1969. I would not necessarily choose a watch like this as my everyday piece, but it has become an icon with its eye-catching square case and pushers on the right side of the chronograph. It was originally made in two different versions, one with a blue dial and white registers and the other with a gray dial and registers. The blue is definitely the most popular amongst collectors, and I choose the blue dial with white indicators and a red second indicator, or "Gulf" version with light blue and orange stripes from their logo.

Steve McQueen in the movie *Le Mans* which featured the legendary Monaco.

BREITLING NAVITIMER

John Travolta has been an ambassador for Breitling for years.

I am particularly fond of Breitling's two models Navitimer and Emergency (the latter is described on pages 144–145). Breitling Navitimer World with a black leather strap is my choice. This mechanical piece was first launched in 2005, with two time zones and a 42-hour power reserve. After its introduction in 1952, Breitling became a fast growing favorite brand amongst pilots. This icon is Breitling's most famous and popular model, still easily recognized by its classic design. Since the well-known classic 806 model appeared in 1952, Breitling has managed to stay true to its characteristics and has kept its soul.

The world's oldest chronograph is still a popular watch with pilots. It has a slide rule for calculations. It calculates ascent and descent rate, conversion of nautical miles to kilometers, etc. Another feature: it can calculate average speed. The case measures 46 mm and is waterproof down to 30 meters. It also has GMT.

Breitling Navitimer with brown leather strap, 46 mm chronograph, and GMT.

BLANCPAIN
FIFTY FATHOMS

BLANCPAIN IS IN MANY WAYS A RATHER UNKNOWN BRAND, BUT DIVERS KNOW BLANCPAIN'S FIFTY FATHOMS VERY WELL.

The model that was introduced in 1953 turned out to be one of the company's greatest successes. The name Fifty Fathoms is in this case reflecting diving depth. How did Blancpain come up with this? The watch almost instantly became an icon and is ranked as one of the top quality diving watches in the business. The watch is often compared to Rolex Submariner, Sea Dweller and Omega PloProf, all classic, recognized diving watches. Like so many great performers, Fifty Fathoms is a product created from genuine passion. We are talking about the passion of Jean-Jacques Fiechter, the director of Blancpain from 1950 to 1980.

Fiechter was a diver himself, and when Blancpain was contacted by the French army, his chance to create a diving watch became reality. It was a great opportunity for Fiechter to combine his two greatest passions.

Even today, the popular timepieces are so much more than diving watches. Men want a robust, sporty watch just as much for wearing with a dark suit as for wearing with jeans. I suppose it is just as common to wear a contemporary diving watch with business dress in a meeting as to wear it underwater.

Following the Second World War, on the initiative of two heroes of the Free French Forces, the French navy created its elite unit of "Combat Swimmers." Maloubier had been a secret agent in the French Special Forces during the war. The divers in the elite group were trained in undersea intelligence and sabotage actions. The group was specially trained in attacks in sea ports, and could perform risky actions against ships and strategic targets. Their tasks were to identify, neutralize, and remove explosives, both conventional and improvised, either sub-marine or on ships. Naturally, most of these actions were performed at night.

The divers were equipped with black diving suits, weapons, gloves, knives, valves, fins, and masks, but Maloubier and Riffaud also acknowledged

the importance of robust and reliable diving instruments. The three most important scuba items were a compass, a depth meter, and a diving watch. The diving watch was central to many of the key tasks divers were confronted with. Timing of the dive was unarguably essential, but timing for navigation purposes might be a less obvious need.

After running tests on the watches available on the market at the time, Maloubier concluded that none of them were up to the task. Thus, he decided to undertake the design of a timing instrument that would target the needs of military combat diving. Maloubier wrote up detailed specifications for his diving watch and farmed them out for bidding. Unfortunately, the reception from the industry was discouraging and one of the commercial directors for the firm LIP even

French combat divers, Suez, 1957.

commented that such a timepiece "would have no future."

For both commercial and administrative reasons, the French Navy decided to buy all their diving equipment, including Blancpain diving watches, through one single source, Spirotechnique.

"Finally a small watch company who understands and would agree to go ahead with our project. A watch that can handle the environment..."

Robert "Rob" Maloubier, when he met Blancpain

It seemed the industry at the time had a different focus area, especially on aviation watches. Known brands such as Fortis, Sinn, Bell&Ross and Breitling are all linked to aviation. Diving watches were, at that time, far from their priority.

But then, the two aforementioned marine officers were officially introduced to Mr. Fiechter of Blancpain. With Fiechter's personal passion and fascination with diving, he needed little convincing. Blancpain quickly agreed to develop the timepiece for the École des Nageurs de Combat (Combat Diving School).

Maloubier describes his first meeting with Blancpain: "Finally a small watch company, who understands and would agree to go ahead with our project. A watch that can handle the environment, with a black dial, bold large numerals and clear markings: triangles, circles, squares; a rotatable exterior bezel which repeated the markings of the dial. We wanted at the start of a dive to be able to set the bezel opposite the large minute hand in order to mark the time. We wanted each of the markings to shine like a star for a shepherd."

Fiechter, an experienced diver, had his own ideas. He found out that the rotating bezel should rotate in one direction only.

This way, the bezel could not be rotated by mistake. Any mistakes would shorten a dive, rather than lengthen it beyond the air supply. Of course, he knew that the watch had to be water-resistant.

Furthermore, he realized that manual winding risked wear on the waterproofing system for the crown, so he decided that the movement would have to feature automatic winding, minimizing the number of times that the crown would need to be pulled out. Finally, he felt that protection from magnetic fields was required for a timepiece that would be used in the harsh conditions of combat.

Combining his specifications with those of Maloubier and Riffaud, Fiechter and his team now had their design goals.

It is remarkable that the set of characteristics which Blancpain and the French Navy jointly developed in 1953 have ever since defined the finest diving watches: high water resistance, robust protected crown systems, automatic winding, black dials with clear luminescent markings, uni-directional rotating bezels with timing markings, and anti-magnetic protection.

This is the anniversary model honoring Fifty Fathoms.

There was one last touch that Blancpain added to the design. At the six o'clock position on the dial, a humidity indicator was added. In the form of a small circle, the indicator showed blue if the air in the case was dry. If water had penetrated, the color would change to pink, as a warning. Blancpain's pioneering effort was named the "Fifty Fathoms" after the British measurement of 50 fathoms, or approximately 91.45 meters, which was considered the maximum depth a diver could achieve with the oxygen mixture then in use.

For both commercial and administrative reasons, the French Navy decided to purchase all their diving equipment, including Blancpain diving watches, through one single source: Spirotechnique, situated on the Left Bank in central Paris. This same business at the time enjoyed a relationship with Jacques Cousteau, the inventor behind the famous diving regulator valve. Cousteau heard of the Blancpain Fifty Fathoms watch, and later selected it for use in the historic dives chronicled in the film *Silent World*, winner of both an Oscar and a Palme d'Or at the Cannes Film Festival in 1956.

Meanwhile, other countries' navies followed the lead of the French in selecting the Blancpain Fifty Fathoms for their divers. The Israeli, Spanish, German, and U.S. navies were all key Blancpain customers.

It was not always easy to achieve this wide military acceptance. For example, the U.S. military placed a variety of obstacles in the way of any non-U.S. watch producer. Under the existing "Buy American Act," American producers were given a 25% price advantage over foreign competitors. The jewels used in the watch had to be sourced from a supplier in Missouri. Fortunately for Blancpain, there was an advocate for the Fifty Fathoms who believed in the watch: Allen Tornek. Tornek had become an acquaintance of Jean-Jacques Fiechter through their common passion for diving.

When he learned of the U.S. military's call for contractors, Tornek immediately thought of Blancpain. However, to meet the requirements of the bidding process, he had to deal with the labyrinth of laws standing in his way, such as the Missouri source for jewels.

Tornek's solution was straightforward; he simply bought the Missouri jewels, and, finding them inferior, threw them away. Tornek won the bidding process, and delivered the watches to the U.S. military under two names: "Blancpain Tornek" and "Rayville Tornek" (Rayville was a name used by Jean-Jacques Fiechter for some of the Blancpain production).

Like many of the military watches, the Tornek series complied with particular military specifications. Indeed, the Tornek series watches bear the label "MIL-SPEC 1" on the dial. One element commonly found in the military specifications dealt with the luminosity of the dial and bezel markings. In the U.S.'s case, as with many other

military units, it was required that the Fifty Fathoms use radioactive material, such as Promethium 147, so that the indications would "glow" in the night conditions envisioned for many dives. These MIL-SPEC materials were fearsome, even by the rather casual standards of the day that applied to radioactive elements. The cases bore an inscription reading "DANGER. IF FOUND RETURN TO NEAREST MILITARY FACILITY."

As the Blancpain Fifty Fathoms established itself as the worldwide standard for military diving watches, it achieved similar success in the civilian arena. At the time, mechanical diving watches were not considered a horological luxury. They were serious instruments and part of the required kit for any diver. So some of Blancpain's historic production of the Fifty Fathoms was sold under the name of the diving shop supplying the rest of the diving equipment.

These shops sold products under the name "Aqualung." "Aqualung" was the name Jacques Cousteau called his line of diving equipment, centered on his pioneering development of the first truly useable scuba air regulator. Cousteau's Aqualung can rightly be credited with giving birth to modern scuba diving. And the diving watches sold in his stores were Blancpain's bearing the label "Aqualung."

Not only were the Blancpain Fifty Fathoms watches the inspiration for all of the diving watches that have copied its list of features, but in some ways they can be seen as trendsetters for the mid-size and large-size watches that are still fashionable.

The style of the 1950s dictated that wristwatches had to be small. It was the norm that men's watches were 32 to 34 mm in diameter. This size, appropriate for a dress watch, did not suit the mission of a serious dive watch. Where legibility was critical, as the saying goes, size matters. So the Fifty Fathoms broke with the fashion of its time and offered large diameters

THE MODELS

There have been many different versions produced since the start of production in 1953. Here is a small sampling.

Watch made for the 50th anniversary in 2010. 45-mm bezel.

Civilian Fifty Fathom model from 1965–70. The bezel has a diameter of 41 mm.

Fifty Fathoms Flyback from 2012. 45-mm bezel.

The first Fifty Fathom model from 1953. The bezel has a diameter of 42 mm.

Fifty Fathoms automatic in brushed gold from 2012. 45-mm bezel.

to improve legibility, but the design also appealed to the fashion industry.

Blancpain has offered a number of varieties of the Fifty Fathoms. They all share the same DNA, adhering to the general specifications laid out by Maloubier and Fiechter in 1953.

The 50th anniversary version of the Fifty Fathoms was launched during the Basel fair in 2003. It was offered in three limited series of 50, and became highly sought-after. For the first time, the bezel was fashioned out of sapphire. Not only did this bring a sheen which no other material could match, but it was practical too because of its scratch resistance.

The movement was updated also, with the use of the caliber 1150 movement offering a 100-hour power reserve. Finally, Blancpain presented an ingenious strap attachment system, making it easy to adjust underwater. So easy that Blancpain's CEO Marc A. Hayek, during a dive with Robert Maloubier launching the watch in Thailand, demonstrated a strap change underwater. All the anniversary pieces sold out in an instant, and today they command a hefty premium on the used market. Over time, more than 50 different varieties of the Fifty Fathoms have been developed.

So the Fifty Fathoms traces a rich lineage over more than half a century. And the passion that gave birth to the line and its uncompromising adherence to the needs of the world's most demanding divers endures.

This is my favorite model. Simple, clean and no nonsense. Blancpain Fifty Fathoms caliber 1315 launched in 2007.

ALBERT

Albert Einstein owned two Longines watches.
One was an elegant rectangular model in gold
from 1929. This watch ended up being the most
expensive Longines ever sold in an auction,
when it was sold in New York in 2008 for close
to $600,000. The other watch had *Albert Einstein*
engraved on the back of the case together with
the date he received the Nobel Prize:
February 16, 1931.

THE WORLD'S FIRST SMART WATCH

WITH ITS CONSTANT LAUNCH OF NEW MODELS AND FUNCTIONS, "I'M WATCH" IS DIFFERENT. BUT IT HAS A REMARKABLE SIMILARITY TO APPLE DESIGNS. THIS IS A WATCH FOR TECH NERDS.

No doubt Apple must have been the inspiration for creating this product. It is intuitive and the user interface is sleek.

The watch can receive calls, SMS texts, e-mails, and allows you to check your calendar and Facebook. It streams music and has several apps already built in. So far, it's possible to use or access Instagram, games, music apps, weather, compass, Twitter, the stock market, your address book, etc. If you receive a phone call, you may leave the phone in your pocket and answer directly from the watch, both over microphone and via headset.

The watch itself has a small TFT touch screen measuring 1.54 inches with 240 × 240 pixels. The case measures 40.6 × 52.9 mm with a height of 10 mm. It contains a 4GB Flash drive and 128 MB of RAM. The producers promise 24-hour battery time with Bluetooth turned on, and 48 hours without. Tests show it needs charging once every 24 hours. Talk time is 3 hours. It is compatible with IOS 4 or higher and Android 4 or higher.

The most affordable model comes in several different colors and is currently priced at about $315, while the luxury model in white gold with diamonds costs an incredible $12,450.

This watch is modern in form, symbols, and design.

Check out the latest features at imsmart.com

And, check out the i'm Watch official trailer.

WATCH-MAKING TRADITIONS IN FLORENCE

IF YOU LOVE ITALY AND ARE INTERESTED IN WATCHES, NEXT TIME YOU VISIT FLORENCE, YOU SHOULD STAY AT THE FABULOUS HOTEL L'OROLOGIO. IT'S PART OF THE HOTEL CHAIN WHYTHEBEST AND EVERY FLOOR IN THE 54-ROOM HOTEL HAS BEEN DEDICATED TO ONE OF THE LUXURY TIMEPIECE BRANDS ROLEX, VACHERON-CONSTANTIN, AND PATEK PHILIPPE.

Several items reveal that the
owner of this hotel is seriously
interested in watches.

An old Rolex sign from a dealer
hangs outside the restaurant.

Hotel L'Orologio is unique and the interior is
inspired by owner Sandro Fratini's outstanding
watch collection. His personal collection of more
than 2,000 vintage timepieces is exhibited all
over the hotel. Large areas of the hotel have
been furnished with an extremely well developed attention to and inspiration from the world
of watches.

A strong personality and unusual approach were
the basic directions taken by the architect Mari-
anna Gagliardi. The library and the bar area
are both based on the old venerable gentlemen's
club. The attention to detail gives the guests
an exciting experience. Her choice of materials—leather and polished bronze, plush textiles,
parchment, dark brass, and mahogany—con-
jures an aura of tobacco. One of many small

More than 2,000 of owner Sandro Fratini's watches are exhibited in the hotel.

interesting details, the water taps in the bathrooms resemble the crowns of a watch. If you appreciate watchmaker traditions and wristwatches, this is definitely the place to stay next time you visit beautiful Florence. The city is known for its long watchmaking traditions and as the birthplace of Anonimo and Panerai, amongst others.

The first watchmaker's shop in Florence was opened in 1840 by Giovanni Panerai. This was the beginning of the production of watches in Florence and of the Panerai company. The shop, on Ponte alle Grazie, was also the city's first watchmaking school and was named *Orologia Svizzera*.

Panerai supplied watches to the Italian Navy and the Italian Department of Defense. The watches were chosen for their durability; they were waterproof and also used by divers in the navy. Panerai has since grown and become a very well-known brand name in the business and very popular with customers. The combination of Italian design and Swiss precision is in my opinion totally perfect.

There are other watchmakers in Florence. One is Gucci, producing expensive, very trendy luxury articles. Their pieces are of high quality, often made in gold with precious stones, and are highly appreciated by both men and women.

EL PRIMERO— THE WORLD'S MOST STUNNING MOVEMENT

AMONG COLLECTORS AND ENTHUSIASTS, EL PRIMERO IS THE BEST KNOWN AND POSSIBLY THE BEST MECHANICAL TIMEPIECE.

El Primero, "the first," is still the most precise series-made chronograph. The secret is its spectacular cadence. While other works in general perform 8 jumps per second, at best, El Primero jumps 10 times per second. The El Primero has constantly evolved with astounding innovations, including dry lube securing an excellent long-term stability, as well as a power reserve that is exceptional for such a high frequency. All Zenith watches are equipped with a proprietary Zenith movement.

There are few brands in the world that can match their exceptional technique and production of a mechanical timepiece. When you choose a Zenith watch, you are guaranteed you own a watch that has been produced from start to finish in Zenith's Le Locle premises. A guarantee needed for a piece that results from 180 different movements, interpreted in 500 variations.

It takes nine months to make a Zenith watch. During this period, there are no fewer than 80 watchmaking professionals, in turn, devoting their talent and expertise to the watch and its components.

CHARLES VERMOT SAVES EL PRIMERO

The story of El Primero could have ended abruptly in 1975. That was when Zenith Radio Corporation, the American owner of the brand for the previous four years, decided to stop making mechanical watches and stick to quartz. The decision was irrevocable, and the owner's plan was to negotiate a scrap price for all the machines, movements, and tools and sell it all. One watchmaker tried to reason with the American headquarters, with no luck. The order was to sell everything connected to the mechanical production—together with a century-long tradition.

A catastrophic decision. Charles Vermot could not just stand by and witness his production equipment being sold or destroyed. Under the threat of losing his job, he started collecting important tools and components. He was a thorough man by nature. Charles Vermot

The incredibly beautiful El Primero Stratos Flyback Striking 10th.

marked, noted, archived, classified, and protected cutters, tools, and machines carefully. He also made a thorough description and recorded the entire production process on a portable computer.

In terms of important milestones in the development of mechanical watches, this story is one of the most important ones and the start of the famous El Primero timepiece. The new beginning takes place barely nine years later. One fine morning, the entire factory, including previously-lost equipment, is up and running. Without this very brave action, Zenith and El Primero would most likely not exist today. In 1980, the price for one production machine alone was more than 40,000 Swiss francs, or $41,500. And when it takes 150 of them to finish an El Primero, it means a total of 7 million Swiss francs (more than $7,265,000.) Zenith's financial, technical, and human investments could have been deleted in an instant.

Several known brands in the watch industry started looking at Zenith as their possible producer. Slowly but surely, the orders kept coming in and were breathing new life to the production. In 2000, LVMH Group took over and decided to reserve El Primero for Zenith's own models.

AWARDS

Zenith has received more than 1,400 different awards and recognitions over more than a century. This is impressive in itself, but even more impressive is that out of all these awards, Zenith received 1,398 first prizes, and thereby accounted for 98.5% of all first prizes ever granted. These numbers reflect recognition of what can only be described as real technical innovations.

9 months' work

20 watchmakers

5,500 operations

50 milling operations on the dial side and

77 other milling operations on the bridge

5–50 operations per component

18 different metals in the

classic version

Calibre El Primero 4052B

El Primero Stratos Flyback with
Zenith's Caliber 405B movement.

INGENUITY AND ENGINEERING FEATS INFLUENCING DEVELOPMENT

One of the brand's latest technical developments is El Primero Striking 10th, a "jumping seconds" chronograph showing for the first time the smallest fraction of time ever measured by a series produced mechanical movement. It allows the reading of a tenth of a second.

Zenith is known for fully mastering all facets of making beautiful, complicated, and technically advanced mechanical timepieces. During the last decades, the brand has invented and produced tourbillons, open models, Power Reserve, "large-date," showing of a tenth of a second, minute repeater, and perpetual calendar, in addition to moon phase and day/night indications, alarms, and the exceptionally complex Zero Gravity tourbillon. Year after year, the story of El Primero has been described through its development, technological breakthroughs, and its refined new functions.

El Primero's heart beats faster. So does yours.

THE MOST WELL-KNOWN EL PRIMERO MOVEMENTS

CALIBER EL PRIMERO 4026

Automatic. Single spring barrel, 50-hour power reserve and COSC certified chronometer
Functions: Hours, minutes, subsidiary seconds, split seconds, chronograph and large date

Diameter	30.5 mm
Height	9.35 mm
Jewels	32
Balance	Glucydur
Frequency	36,000 vph
Balance spring	Self-compensating flat spring
Shock protection	Kif

CALIBER EL PRIMERO 4047

Automatic. Single spring barrel, 50-hour power reserve
Functions: Hours, minutes, day/night indicator, display of sun and moon phases, chronograph and large date

Diameter	30.5 mm
Height	9.05 mm
Jewels	41
Balance	Glucydur
Frequency	36,000 vph
Balance spring	Self-compensating flat spring
Shock protection	Kif

CALIBER EL PRIMERO 405

Automatic. Single spring barrel, 50-hour power reserve
Functions: Hours, minutes, subsidiary seconds, flyback chronograph and date

Diameter	30.5 mm
Height	6.6 mm
Jewels	31
Balance	Glucydur
Frequency	36,000 vph
Balance spring	Self-compensating flat spring
Shock protection	Kif

CALIBER EL PRIMERO 4052B

Mechanical with automatic winding and approx. 50-hour power reserve
Functions: Hours, minutes, subsidiary seconds, chronograph (10-second totalizer at center, 60-second totalizer at 3 o'clock, minute totalizer at 6 o'clock) and date

Diameter	30 mm
Height	6.6 mm
Jewels	31
Balance	Glucydur
Frequency	36,000 vph
Balance spring	Self-compensating flat spring
Shock protection	Kif

CALIBER EL PRIMERO 4057

Automatic. Single spring barrel, 50-hour power reserve
Functions: Hours, minutes, subsidiary seconds, chronograph shows 1/10th of a second thanks to a fast drive of chronograph hands (1 revolution every 10 seconds) totalizer for 6 revolutions and date

Diameter	30.5 mm
Height	6.6 mm
Jewels	31
Balance	Glucydur
Frequency	36,000 vph
Balance spring	Self-compensating flat spring
Shock protection	Kif

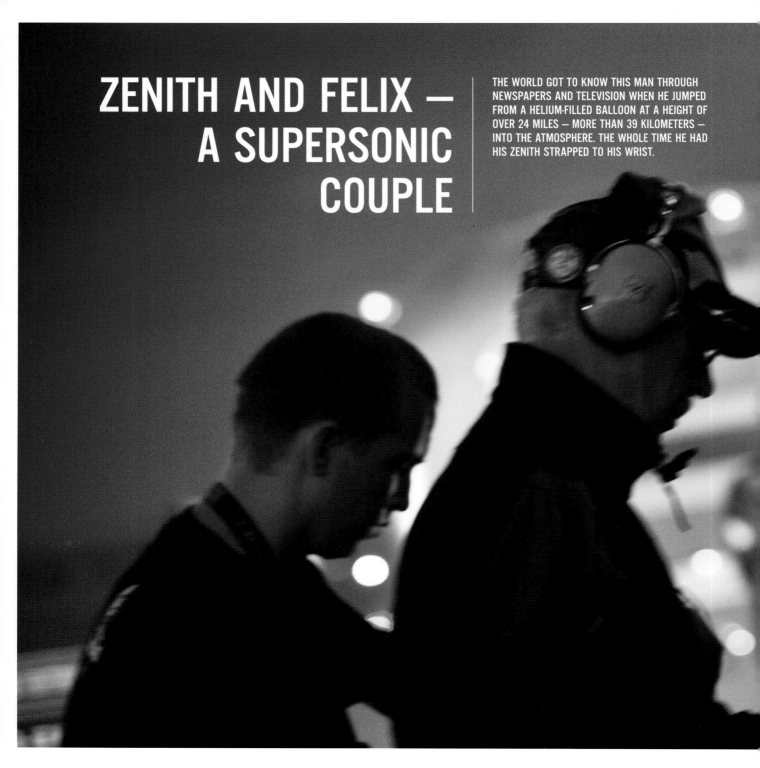

ZENITH AND FELIX — A SUPERSONIC COUPLE

THE WORLD GOT TO KNOW THIS MAN THROUGH NEWSPAPERS AND TELEVISION WHEN HE JUMPED FROM A HELIUM-FILLED BALLOON AT A HEIGHT OF OVER 24 MILES — MORE THAN 39 KILOMETERS — INTO THE ATMOSPHERE. THE WHOLE TIME HE HAD HIS ZENITH STRAPPED TO HIS WRIST.

The world heard of this man through the media. On October 14, 2012, he launched himself from a balloon at an altitude of 24.2 miles (39,045 meters), four times higher than airliners fly. This attempt to challenge the atmosphere fulfills a lifelong dream for the adventurer.

Felix "Fearless" Baumgartner was born in Salzburg, Austria, in 1969. He started skydiving at the age of 16 and continued perfecting his skills in the Austrian military, and later on competition teams. In 1988 he started performing parachute jumps at events organized by energy drink producer Red Bull. The company's "out of the box" thinking and Baumgartner's adventurous nature made a good match, and they have worked together ever since.

By the end of the 1990s, Felix Baumgartner felt he had reached as far as was possible within traditional parachute jumping. He therefore

This is the watch Felix Baumgartner was wearing on his mission into the atmosphere: Zenith El Primero Stratos Flyback with a steel bracelet. Baumgartner is also an ambassador for the famous and historical brand.

started BASE jumping. This means you jump from fixed objects like a building, antenna, span, or a mountain. He discovered that the quick reflexes and precise techniques needed for these low jumps also strengthen the technique used in parachute jumps of extremely high altitude.

Baumgartner has completed famous BASE jumps and has been nominated for the World Sports Award and in two categories in the NEA Extreme Sports Awards. He is also a spokesperson for the organization Wings for Life Spinal Cord Research Foundation and is a helicopter pilot. He realized that the Red Bull Stratos Mission was a step into the unknown, taking on the challenging project of breaking the sound barrier without vehicular power. The result of this stunt could lead to important research data—both medical and scientific.

It was decided: With a team of experts, Felix Baumgartner would perform a jump from an altitude of 120,000 feet (36,576 meters) in a capsule inside a helium balloon.

The Red Bull Stratos team had world experts within aviation, medicine, and engineering, including experts in development of pressure suits, space capsules, and balloon production. The team included the now-retired Joseph Kittinger from the United States Air Force Academy, who has ascended to an altitude of 102,800 feet (31,333 meters), and who held three of the records Baumgartner would be trying to break.

The project's goal was to break human records existing for over 50 years, and to attempt to break these four world records: the highest manned balloon flight, greatest free fall, longest lasting free fall, and (for the first time in history) becoming the first person to break the sound barrier.

As preparation for the record attempt, on March 15, 2012, he completed a test jump from a balloon from 71,615 feet (21,828 m) and Baumgartner thus became the third person in history to safely parachute from such a height.

Baumgartner's free fall was seen by people all over the world and he broke the speed of sound with Mach 1.124 during his 4 minute and 20 second long jump to the ground. He is the only person in history who has performed this without vehicular power.

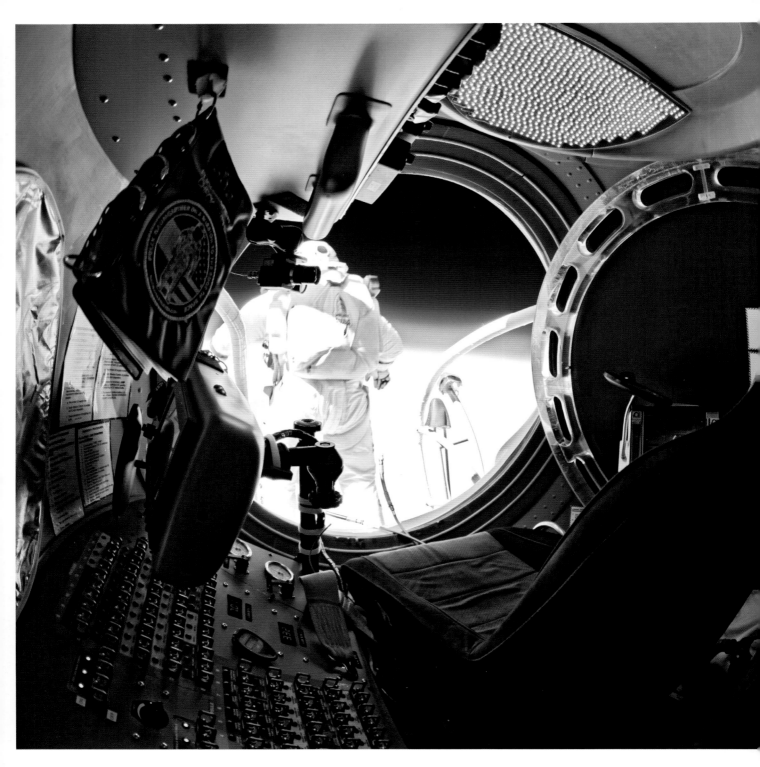

PHYSIOLOGICAL EFFECTS

Unprotected exposure in this area of the stratosphere can, without aggressive medical treatment, kill a human being in a matter of seconds. There is not a complete vacuum, but the air molecules are so few that it is impossible to survive without a protective suit or capsule.

So far no medical treatment has been developed to handle long term exposure of these hostile elements. Red Bull Stratos Team made up their own procedure in case of emergency. Baumgartner's suit and capsule were designed especially to minimize the risk at such an altitude.

THE SPEED OF LIGHT

Baumgartner would be flying through a lack of air density, so would reach the speed of sound at about 100,000 feet, starting at 120,000 feet; Mach 1 (speed of sound) is about 690 mph or 1,110 km/h. If he continued through the speed of sound, he would be supersonic.

Red Bull Stratos Team agreed that the planned height for Baumgartner's jump was a key factor for reaching their goal. The same thin atmosphere that represents many physiological dangers is also Baumgartner's ally when supersonic. Thin air represents almost no resistance; hopefully, it would minimize the effect of the shockwaves. Even if Baumgartner moved in only 1% of the earth's atmosphere, gravity is still present. He would fall in an accelerating speed up until the point where air density starts to increase—he wouldn't fall any faster; in fact, the descent rate would decrease. As he approaches the ground, air molecules increase and the temperature rises, so he'd be at the lowest speed just before the parachute opens at around 5,000 feet.

Felix Baumgartner has completed world known BASE jumps and has been nominated for a World Sports Award and in two categories of NEA Extreme Sports Awards. He is also a helicopter pilot.

THE ATMOSPHERE

Normally, the atmosphere gradually becomes colder as one ascends. At the altitude of an airplane at about 35,000 feet, the temperature drops dramatically to –50 or –70° F.

The cooling or fall in temperature stops at the beginning of the stratosphere, which by the way is never at an exact height. The stratosphere is exactly like all other layers of the atmosphere: it starts and ends at different heights, depending on where you are on earth. These layers are dependent on the temperature, so near the equator, the height of the stratosphere will vary from its height near the North Pole. These variations also change with the seasons. So, the reason the temperature actually increases when you pass through the stratosphere is due to the ozone layer. Ozone molecules absorb energy and feed heat back to the stratosphere. But it is still cold: 0 to 10° F.

SCIENTIFIC VALUE

The goal for Red Bull Stratos was to provide data for the development of space safety, develop a new generation of space suits—including increased mobility and visual clarity—and assist in developing systems, protocols, and routines in order to get crew who are exposed to extreme heights and acceleration safely back from space.

Another important goal was to assist research on the effects on the human body during supersonic acceleration and retardation, which is closely connected to the development of the latest innovations within parachute systems.

The weather was constantly changing, and the Red Bull Stratos team had to adapt. It meant the entire mission could be moved and that even during the mission, it could be necessary to adjust calls, schedules, and replacement of equipment. All testing prior to the mission was based on the ability to change all plans and details according to weather.

To comply with the regulations of the Federal Aviation Administration (FAA), the Red Bull Stratos balloon would not be able to fly if the sky was more than $^5/_{10}$ (half) clouded, or if the horizontal view at any time was less than 3 miles (4.8 km). Clear skies meant better tracking of the capsule's ascent and Baumgartner's jump.

On October 14, 2012, Felix "Fearless" Baumgartner did his supersonic freefall over the New Mexico desert. The mission became an instant success and Baumgartner described the feeling of breaking the speed of sound as "like swimming without touching water."

Baumgartner's mission was watched by people all over the globe when he broke the speed of sound with 1.24 Mach during his 4-minute-20-second freefall to the ground. He is the only person who has done this without a supersonic plane or space shuttle.

Felix Baumgartner jumped out of a capsule suspended from a balloon in the stratosphere at an altitude of 24.261 miles (more than 39 kilometers). The mission showed that it is possible to exceed human limits dating back more than 50 years, and in addition Baumgartner broke four world records.

At one point during his freefall Baumgartner appeared to spin rapidly, but he quickly regained control, and moments later opened his parachute and landed perfectly on his feet in the New Mexico desert.

A fact not too many people know is that Baumgartner was wearing the brand new El Primero Stratos Flyback Striking 10th chronograph from Zenith and that he is one of the foremost ambassadors for this prestigious watch brand. Zenith is one of his most important sponsors and his only supplier of watches and time functions.

El Primero Stratos Flyback Striking 10th with black croco strap.

SPACEMASTER
Z-33

Spacemaster Z-33 is unlike any other watch and should not be confused with other traditional pilot watches. This one has everything the others have and more. The bezel is massive and some find it too big. The watch has strong luminescent skeletonized hands showing hours and minutes floating above the digital dial. The black LCD screen with red segments is illuminated by the LED backlight, which adapts automatically to ambient luminosity: the digital displays' transreflective technology takes into account ambient light, allowing the digital data to be read as easily on night flights as in direct sunlight. This technology also helps conserve battery power and is distinct enough not to interfere with the digital information.

Omega Spacemaster Z-33 is an unusual and special watch, and has its fans among watch enthusiasts.

The watch has two time zones with either 12- or 24-hour display, an alarm, and a perpetual calendar, and of course chronograph-function with a countdown timer. It can log up to ten flights and visualize those logs with date-hour indications. If you wish to read the display, simply press the pusher at the 8 o'clock position to disengage the hands.

The case is made of brushed titanium, is fitted with a scratch-resistant sapphire crystal with anti-reflective treatment, and is water resistant down to 30 meters.

CARS AND WATCHES

BEAUTIFUL AND EXCITING WRISTWATCHES HAVE BEEN LINKED TO FAST SPORTS CARS FOR A LONG TIME NOW. THEY HAVE A LOT OF COMMON TRAITS THAT ATTRACT EMOTIONS AND INTEREST. WATCH MANUFACTURERS UNDERSTAND HOW THIS ENGAGES MEN.

It is no coincidence that IWC chose to launch their Ingenieur series at the same time Mercedes launched their AMG beauties. These are elements that make grown men lose their common sense. "Engineered for men," IWC called it. IWC and Mercedes together go a long way back. IWC makes the AMG model watch in Mercedes' dashboard. Porsche Design used to make IWC watches, but has lately been tied to Eterna.

Porsche has had its natural connection to Porsche Design studio. The company no longer has any connection to the Porsche cars, but they design pens, bikes, watches, bags, sunglasses and other products, together with high-end merchandise. Porsche Design is independent with their watches Dashboard and Flat Six (which obviously references Porsche's 6-cylinder boxer engine), Worldtimer and Indicator. The Indicator model was especially made for Porsche's supercar project: the immensely beautiful and legendary Porsche Carrera GT, built as a full-blooded race car in composite materials with a 612 HP V10 engine. My absolute dream car. The movement is a modified ETA-work with more than 800 different parts from Porsche Design. The case is 48 mm thick and made in titanium. The buttons on the watch are copies of the pedals in the car, and the wheel itself in the mechanical works is made to resemble the custom-built wheels. It is also a mechanical wonder with very complicated functions. They make only two to four pieces every year, and it has, of course, a waiting list. It is, like the car, not exactly cheap...from $181,300 to $378,000.

I was able to get hold of a Motochron watch a few years ago. The watch is made as the speedometer of the Porsche 911. At the time I owned a Porsche 911 and felt I just had to have the watch. Motochron has made several models with inspiration from different car instruments.

Porsche's beautiful Indicator watch that was launched together with the Carrera GT model.

"You may not be able to buy the car, but we can help you get some of the same feeling."

Watches are often tied to legendary car races. Chopard has its own models linked to the well-known Mille Miglia ("thousand miles") race. This classic Italian car race, originally run between 1927 and 1957, today has become a race of vintage cars from that time period. Beautiful cars from Alfa Romeo, Porsche, and Ferrari compete in the same category as Targa Florio, Grand Turismo, and Carrera Panamericana. The rubber strap on the Chopard watch has a car-tire pattern and the watch has many other exquisite details. The Alpina watch company, from Geneva, has its own series for the twelve-hour-long Sebring race. The race takes place on Sebring International Raceway in Florida...the track where the Danish racer Tom Kristensen has record wins.

Blancpain is the sponsor for Lamborghini
drivers in the motorsport series.

He has also won the world's oldest active sports car race the most times—Le Mans, also known as the grand prix of endurance and efficiency. Kristensen has eight Le Mans victories and is simply addressed "Mr. Le Mans." I had the opportunity to meet him during Le Mans a few years ago—brilliant!

Tom Kristensen has his name on a watch from Rosendahl. The watch is a tribute to Denmark's most famous sports personality of all time.

The British watch manufacturer Graham worked with Mercedes' GP Silverstone series. Audi has been less visible with their watches from Tachoscope. Dutch sportscar manufacturer Spyker, not particularly well known, makes an automatic rose gold chronograph: a handmade, exclusively-numbered watch meant for buyers of their cars, and ranging from $6,300 to $21,600. Tonini Lamborghini is offering his customers characteristic and unique timepieces: large, massive wristwatches with an almost triangular shape to the dial and with the Lamborghini logo clearly placed on the bezel.

One of the most familiar cooperative ventures is that of Bentley and Breitling, named Breitling for Bentley. These are beautiful and expensive watches launched with Bentley's different models, for example, a Tourbillon for Bentley's Mulliner.

The watch can be removed from the dashboard and be used as a wristwatch.

Breitling for Bentley GMT.

This has turned out to be a great success for Bentley, but particularly for Breitling. Many people are dreaming about a Bentley but have to settle for a Breitling.

Other smaller watch series are tied to a single car model. Italian Alfa Romeo is linked to Italian Giuliano Mazzuoli. Hublot made its own watch with the launch of Morgan's Aero 8 which, entirely appropriately, was named Hublot Aero Big Bang. This was the only one they made. Aston Martin has a custom built watch, the AMVOX Transponder, and Girrard Perregaux's model is linked with Rolls Royce. Nissan's luxury brand, Infinity, made with Bell&Ross, has two models mainly made for the American and Asian markets. Tag Heuer and McLaren worked together on the watch MP4.12C.

Richard Mille, for years the main partner and the official timer for the Le Mans Classic race, also produces his own Le Mans series of timepieces. Most watch manufacturers recognize the value of being connected to an attractive car brand or some form of motor sport.

Breitling for Bentley model 6.75 with Breitling 448 works. Diameter 48.7 mm; 42-hour power reserve.

BENTLEY AND BREITLING

The famous cooperation between Bentley and Breitling has its own series named Breitling for Bentley. Launched with Bentley's different models, the series includes beautiful and expensive watches for Continental GT and a Tourbillon for Mulliner. This has been a great success for Bentley, but particularly for Breitling.

AUDI AND TACHOSCOPE

The legendary
Auto Union Type D.

When Audi marked its 100-year anniversary, it chose to collaborate with Chronoswiss and made a limited-production series with special features. The limited production of 100 timepieces was made in white gold and platinum. It has a tachometer, stop watch function, and separate timers for seconds and hours. The overall design is inspired by Audi, with the Audi rings placed centrally on the bezel. The watch was launched with a price tag of around $20,000 for the white gold version and $33,000 for the platinum. Like many other car brands, Audi also offers a series of watches engraved with the Audi rings at a more affordable price level.

Parmigiani's Bugatti Super Sport — 10-day power reserve, water-resistant to 10 ATM and in white gold. Limited edition $260,000

BUGATTI AND PARMIGIANI

Bugatti Super Sport in white gold was unveiled at Pebble Beach, August 2010, and became an instant hit. Two years later, Parmigiani Fleurier launched the much-awaited new model in rose gold. The watch is a close collaboration with car producer Bugatti and the model Veyron EB16.4. This collaboration has resulted in several models with the same unmistakable design. Parmigiani supplied the clock in the car's dashboard as well.

I simply must talk about the car, also. Bugatti Veyron has long held the record for the world's fastest series-produced car with a speed of 408.47 km/h and it was voted the car of the decade by TopGear. With four-wheel drive and a 1200 hp, 16-cylinder, 8-liter engine with no less than 4 turbos, this car is in a class of its own.

Excitement abounds when Parmigiani, known for their quality timepieces, launches watches with Bugatti with designs reflecting the Veyron.

The manually wound Calibre Parmigiani 372, with a power reserve of 10 days, was designed to match the new Bugatti. On the wrist, the watch is placed at a 90-degree angle, while the case and power reserve are situated on top of the watch. Several elements are clearly inspired by the car.

Beautiful and aesthetic parts are cut and formed to mimic different parts of the Bugatti Veyron. The company has, in the best tradition, made an exquisite, extraordinary watch with six sapphire crystals and 333 components. It is waterproof down to 10 meters. The dial is made of black

The unique and beautiful Bugatti Veyron EB 16.4 is registered as the world's fastest series-man-ufactured car so far. Many have tried beating this record, but no one has succeeded yet.

opaline with a gold base, a testimony to the Bugatti Veyron 16.4 Super Sport, with transparent sections in the center allowing a glimpse of the hour wheel.

The exceptionally slender profile of the Bugatti Super Sport may evoke the fuselage of a wing, but was in fact designed to reference the emblematic form of the car's wings. It is ergonomic on the wrist, but requires some getting used to.

Graham Mercedes GP Silverstone. Caliber G1735, automatic movement with chronograph, sapphire crystal, transparent case back.

GRAHAM AND MERCEDES

The British watch manufacturer Graham has worked with Mercedes' GP Silverstone. Graham launched "Mercedes GP" watches.

Panerai Ferrari Grandturismo GMT
Automatic, with black croco band.

FERRARI AND PANERAI

I really like Panerai and found it very exciting that the two famous Italian brands Panerai and Ferrari were to work together. After a few years, Hublot took over as Ferrari's partner in 2011.

Hublot Ferrari Big Bang Magic Gold. The case is 45.5 mm. With sapphire glass, gold, and a black crocodile strap. Produced in a limited edition of 1,000.

HUBLOT AND FERRARI

In 2001, Hublot took over as partner with Ferrari. This partnership is closely related to motor sport and F1 with "Official Watch" by Ferrari and Scuderia Ferrari, "Official Timekeeper" by Ferrari, Scuderia Ferrari, and Ferrari Challenge. The deal was solemnly announced by Jean-Claude Biver, Managing Director at Hublot, and Luca Cordero di Montezemolo, president of Ferrari SpA, at the Mugello International Circuit near Florence.

The beautiful new F12 Ferrari Berlinetta with 12 cylinders and the most powerful engine ever built.

This collaboration has proved to be a successful one. Hublot has launched several models with HUB1241 Unico Automatic, Big Bang Titanium and Magic Gold—the famous first 18-karat gold scratch-resistant watch...carbon bezel and black strap with red stitching.

Other models are aptly named Big Bang Ferrari Red Magic Carbon, Big Bang Ferrari King Gold Carbon, and Big Bang Ferrari All Black. All are 45 mm in diameter. They are designed in Hublot style with carbon case and anti-reflective sapphire glass. They are big and sturdy with unmistakable Big Bang elements: the shape of the case, elements on the face, and the Hublot logo on the seconds hand. Ferrari is known for putting its logo on anything and everything, and this watch is no exception, but features a more sober, discreetly-placed logo.

The model has a practical flyback chronograph with visible components. It is a complicated watch, with around 330 different parts and power reserve of 72 hours.

The partnership has resulted in models with Hublot's Chronograph Tourbillon in a limited series of twenty pieces. The King Power series from Hublot also has a license agreement with F1.

HUBLOT AND MORGAN

For the launch of Morgan's Aero 8, Hublot made a watch which suitably got the name Hublot Aero Big Bang. The only one made.

Bell&Ross model BR03-92S Commando in a limited edition.

INFINITY AND BELL&ROSS

The relatively unfamiliar Infinity,
which is Nissan's luxury brand,
worked together with Bell&Ross.
Although this never became a hit,
it still resulted in a watch model
carrying the Infinity name.

Jaeger LeCoultre JLC Aston Martin AWVOX2 chronograph in a limited edition of 200.

ASTON MARTIN AND JAEGER LECOULTRE

Aston Martin DBS sleek driver environment. The partnership with LeCoultre has resulted in models with the company's "Transponder" technique, enabling you to start the car with the watch.

While some of us dream about driving an Aston Martin or buying a Jaeger LeCoultre watch, Aston Martin owners can now have a little taste of James Bond. Jaeger LeCoultre has developed a unique timepiece together with Aston Martin, solely for the owners of the DBS. If you are the owner of the flagship Aston Martin DBS with 510 hp ($330,000), you may buy the reasonably-priced watch for $35,000. This watch is exclusively for the owners of this car and has been made in an edition of only 200. But if you already have the car, this watch can unlock it and start the engine. The watch consists of more than 400 parts and, interestingly enough, engineers shrunk the DBS transponder electronics and placed them in the watch. On top of this, Jaeger LeCoultre also launched a model in rose gold and titanium.

LAMBORGHINI

Tonini Lamborghini offers its customers characteristic and special watches, unlike any others...large watches in an almost triangular form, with the Lamborghini logo in clear view.

ROLLS ROYCE
AND GIRARD PERREGAUX

The special feature of this watch
is that you can both wear it on
your wrist and position it in
your car.

PORSCHE AND MOTOCHRON

Motochron has several models
inspired by different car
instruments. "You may not be able
to buy the car, but we can help
you get some of the same feeling."

Porsche Design P6750 Worldtimer. Automatic Eterna 6037 in titanium with a rubber strap.

PORSCHE AND PORSCHE DESIGN

Porsche Design Studio and Porsche is of course a natural match. Even if they are now two separate companies, there is still a strong relationship via their names. Porsche Design manufactures a whole line of other products like bikes, travel bags, pens, and clothing.

ROSENDAHL AND TOM KRISTENSEN

Tom Kristensen: the driver who has won the most Le Mans races and is simply called "Mr. Le Mans." I had the opportunity to meet him during Le Mans a few years ago. Tom Kristensen, who is Danish, has his name on a watch from Danish Rosendahl. The watch is a tribute to Denmark's most famous sportsperson of all time.

BRUVIK –
A PIECE
OF NORWAY

AS A FIFTEEN-YEAR-OLD, RUNE BRUVIK TRAVELED FROM NORWAY TO SWITZERLAND WITH HIS FAMILY. THEY BORROWED A CABIN IN THE ALPS THAT BELONGED TO A FORMER COWORKER OF HIS FATHER. IN GENEVA, STOPPING ON THEIR WAY TO PICK UP THE KEYS, BRUVIK GOT TO CHOOSE A SWISS SWATCH FROM THE OWNER'S SON'S COLLECTION. HE CHOSE AN ALL BLACK ONE WITH ORANGE DETAILS. THIS WATCH BECAME THE START OF A GREAT PASSION FOR TICKING WATCHES AND THEIR DESIGN.

In the years to come, watch design became Bruvik's main hobby, and where others choose to spend as little time as possible on airport transfers, Bruvik made sure he always had plenty of time for browsing through watch shops.

Bruvik did not study watchmaking, but chose physiotherapy as his profession. But his passion for wristwatches and his dream to design his own never faded. He made design sketches, and had plans for different concepts that would fit his style.

In the summer of 2008 Bruvik decided to search for a manufacturer in Switzerland, to turn his seventeen-year-long dream of designing his own watch into reality.

It was challenging to find a manufacturer in Switzerland who was at all willing to look at the possibilities for developing/prototyping a model. Bruvik contacted the Swiss watch association, who in turn contacted their members. Finally, a manufacturer in Neuchatel was chosen, and during the summer of 2008, the sketches were delivered and things were moving along. The first model was made in a black and white highly technological ceramic material. The material is made under extreme heat and then undergoes a quick cooling process. It is a meticulous and costly process.

This first model/prototype was inspired by arctic Norway, with contrasts from the northern light and glaciers. The BRUVIK black/white watch for men was made in high technology ceramic material, with ETA's automatic 2895-2.

Rune Bruvik is a great guy and obviously energetic about his own watch brand with inspiration from Norway.

Very few manufacturers in 2008 and 2009 were using automatic works in ceramic cases, but it is more common now.

In 2009, the watch company Bruvik Time AS was established and the production of the very first model, BRUVIK black and BRUVIK white, was launched for sale in seven leading stores in Norway. The price for this model was around $1,995. Against all odds, it started to sell really well and allowed the possibility to develop the concept further. The impossible became possible.

Late in 2009, the women's version of black and white became available in a feminine design with diamonds. This was also well received in the market, and in 2010, BRUVIK black gold was launched. This was an automatic Valjoux 7750 chronograph model with matte ceramic case.

Already by 2010, BRUVIK Time had its own stand at the world's largest watch show, Baselworld in Switzerland. And its 90 × 90 centimeter showcase showed there was a world of difference between this young company and the large brands with their 1,000-square-meter stands in the biggest halls. The need to make the new and unknown brand stand out quickly became obvious. Up to that point, the models had been inspired by the natural beauty of Norway, but Bruvik wanted to implement

elements from Norwegian nature into the watches as a completely new and innovative watchmaking style. This resulted in Fjord, an important concept model for BRUVIK.

This model contains Norwegian fjord water from a large waterfall near the west coast of Norway, Langfossen. At the top of Langfossen, 2,007 feet (612 m) above the fjord, the designer collected the water which later was included in a special water capsule at the back of the watch.

Fjord was an important contribution to the collection, and received international interest during Baselworld 2012.

Inspiration from Norwegian mountains and fjords.

A small, exciting company launching a new watch brand with inspiration and elements from Norway and its nature.

Svalbard is available in both automatic and quartz. 45 mm steel case, double domed sapphire crystal front, water resistant to 300 m, screw-in crown, unidirectional turning bezel with superluminova, case back with etched map of Svalbard.

IT TAKES COURAGE AND PERSEVERANCE

IT TAKES COURAGE AND PERSEVERANCE TO ESTABLISH A NEW BRAND IN THIS INDUSTRY. THERE ARE HUNDREDS OF COMPANIES OFFERING HIGH QUALITY WATCHES AND AWARD-WINNING DESIGN. SEA-GOD IS, IN MY OPINION, THE BEST OF THEM ALL.

With Italian design and Swiss tradition, you have a timepiece inspired by the ocean. This was the starting point for making Sea-God.

The Swiss company was established in 1996 and Sea-God was launched in 2003. The entrepreneur and visionary behind it is Enzo Palazzolo. He is a perfectionist and a determined Italian businessman who has had a passion for watchmaking since the early 1980s.

I came across this brand a few years ago. At the time it was just in design prototypes, but I immediately liked the concept and quickly decided I had to have a watch like this. In 2012 when this fabulous watch brand was launched, I was lucky enough to get to know the entrepreneurs, and I bought the Sea-God GMT III.

The logo and symbols are inspired by the Greek sea god Poseidon. All watches in the Sea-God collection can handle dives down to 300 meters.

Sea-God's mission is to create designs mixing traditional mechanics with high quality components, to express the owner's passion for the ocean and luxury.

Sea-God Portorotondo Black Shadow GMT model. All models are waterproof to a minimum of 300 meters.

OMEGA ON THE MOON

THE FAMOUS OMEGA SPEEDMASTER MOONWATCH WAS LAUNCHED IN 1957 AND WAS CHOSEN BY NASA AS THE OFFICIAL WATCH FOR APOLLO. THIS WAS THE WATCH THAT BECAME THE FIRST TO REACH THE MOON, AND LATER WAS NICKNAMED "MOONWATCH." IT BECAME ONE OF THE MOST WELL-KNOWN WATCHES IN HISTORY.

To be able to launch the first watch on the moon, the Omega Speedmaster had to pass a number of tests at NASA. It went through tortures of extreme temperatures, vacuum, intense moisture, corrosion, acceleration, pressure, vibration, and noise. Other brands like Rolex, Breitling, Bulova, Longines, and TAG Heuer all failed the same tests. Omega Speedmaster was chosen to go to the moon.

The story involves the 1960s, and the Cold War between the United States and the Soviet Union. The race to be first on the moon started on October 4, 1957, when the former Soviet Union launched the first satellite, Sputnik. President Eisenhower formed the National Aeronautics and Space Administration (NASA) in July 1958 with the goal to land a manned spaceship on the moon.

The three space programs Mercury, Gemini, and Apollo were each individually designed to bring the U.S. closer to that goal of a successful manned space mission to the moon. Mercury launched one astronaut per mission, the Gemini ship carried two astronauts with each flight, and Apollo's mission was manned with three astronauts.

The space program was an exciting period in America's history and the success of the Apollo mission is without a doubt one of the most incredible human achievements. The Apollo program started in 1967, and by the end of 1972, twenty-four men had traveled from earth to the moon, and twelve of them actually walked on the lunar surface.

The first Project Gemini space flights started in early 1965 and were to include work outside the spacecraft. The first space walk was planned to be from Gemini on June 4, 1965. Each of the three space projects required special equipment to complete their missions, but when the step-by-step procedures of Gemini became clear, the NASA team realized they were missing the specifics for specialized personal timekeep-

ing devices for the astronauts: NASA did not have an approved space watch. Quartz watches were not released before 1970, and good, reliable mechanical waterproof and shockproof watches were hard to find in 1960. The procedures involving designing, manufacturing, testing, and choosing a space-proof watch were time consuming—and NASA did not have much time. So they sent two system engineers to Houston "incognito" to buy several recognized

Omega's classic Speedmaster "Moonwatch" is a must in any watch collection.

chronographs to test for possible use in space. Most astronauts were already experienced and were familiar with the watch industry and specialized hardware. System engineers interviewed the astronauts to find out what kind of watches they were using on their space flights.

Even if it seemed not very likely, NASA engineers hoped to find a few watches able to pass the tests for use in these extreme conditions.

Temperatures on the moon surface vary between –256° and 248° F (–160° and 120° C).

Five different chronograph wristwatches were purchased and taken back to NASA for testing. NASA ordered a number of models of each brand, including twelve Omega Speedmasters, and set up a range of tests to evaluate their use in space. The tests were conducted on March 1, 1965. After the first round, only one watch, the Omega Speedmaster Professional, had passed without serious deviations. Omega Speedmaster was chosen as the official watch for all manned space missions by NASA.

There was only oxygen enough to explore the moon for a few hours.

The Omega Watch Company was totally unaware of these activities, and didn't yet know their watches had been tested and chosen by NASA as the official watch of the American space program.

NASA equipped the astronauts with Omega Speedmasters. The Omegas were first given to the Gemini-Titan III crew. With Grissom and Young, Omega Speedmaster Professional became a part of the standard astronaut equipment. The watch was attached to the space suit with a long black Velcro strip. Onboard Gemini IV, Edward White left the capsule with an Omega Speedmaster on his wrist as the first American to walk in space. Omega Speedmaster was incredibly popular among astronauts, worn just as much in their spare time as on missions in space.

The biggest moment in space program history (and for Omega Speedmaster) was July 21, 1969, when Apollo II landed on the moon. Again, Omega Speedmaster was a part of history and the first wristwatch to be used on the moon. Omega Speedmaster repeated history in April 1970, when it was used to bring the damaged Apollo 13 back to earth. Apollo 13 Lunar Model's power systems were deactivated and the onboard timekeeping devices were not working. Omega Chronograph wristwatches were famously used to precisely time the critical burners that allowed for the safe return of the spacecraft.

The last manned lunar landing planned by Apollo 17 was in December 1972. Bulova Watch Company was anxiously watching this date approach, as they desperately wanted their own watches to be a part of this historic trip.

A letter was sent to the President's advisor, in which Bulova expressed its dissatisfaction with NASA using Swiss chronographs in the American space program.

HIGH TEMPERATURE

48 hours at a temperature of 160° F (71° C) followed by 30 minutes at 200° F (93° C). This under a pressure of 5.5 psia (0.35 atm) and relative humidity not exceeding 15%.

LOW TEMPERATURE

Four hours at a temperature of 0°F (-18°C).

TEMPERATURE-PRESSURE

Chamber pressure maximum of 1.47 x 10-5 psia (10-6 atm) with temperature raised to 160°F (71°C). The temperature shall then be lowered to 0°F (-18°C) in 45 minutes and raised again to 160°F in 45 minutes. Fifteen more such cycles shall be completed.

DECOMPRESSION

Ninety minutes in a vacuum of 1.47 x 10-5 (10-6 atm) at a temperature of 160°F (71°C) and 30 minutes at 200°F (93°C).

ACCELERATION

The equipment shall be accelerated linearly from 1 G to 7.25 Gs within 333 seconds, along an axis parallel to the longitudinal spacecraft axis.

ACOUSTIC NOISE

130 db over a frequency range of 40 to 10,000 Hz, duration 30 minutes.

THESE WERE SOME OF THE STRICT TESTS OMEGA SPEEDMASTER WAS PUT THROUGH.
(Test data descriptions are from NASA)

VIBRATION

Three cycles of 30 minutes (lateral, horizontal, vertical), the frequency of varying from 5 to 2,000 cps and back to 5 cps in 15 minutes. Average acceleration per impulse must be at least 8.8 Gs.

OXYGEN ATMOSPHERE

The test item shall be placed in an atmosphere of 100% oxygen at a pressure of 5.5 psia (0.35 atm) for 48 hours. Performance outside of specification, tolerance, visible burning, creation of toxic gases, noxious odors, or deterioration of seals or lubricants shall constitute failure to pass this test. The ambient temperature shall be maintained at 160°F (71°C).

SHOCK

Six shocks of 40 Gs, each 11 milliseconds in duration, in six different directions.

RELATIVE HUMIDITY

A total time of 240 hours at temperatures varying between 68°F and 160°F (20°C and 71°C) in a relative humidity of at least 95%. The steam used must have a pH value between 6.5 and 7.5.

HIGH PRESSURE

The equipment to be subjected to a pressure of 23.5 psia (1.6 atm) for a minimum period of one hour.

The NASA administration decided that, if a Bulova chronograph was found suitable, it would be used on the last Apollo mission. The astronauts replied that if they were forced to wear a Bulova watch, they would bring an Omega as insurance. Bulova insisted NASA should follow the "Buy American Act," established by the Senate.

Both Omega and Bulova wanted to meet this requirement, but in the end, Bulova never delivered an American-manufactured chronograph.

In August 1972, 16 different companies were notified by MSC (NASA's department for manned space craft) that they planned to establish a QPL (qualified product list) for possible astronaut watches. This list included Breitling Elmore Watch Company, Elgin National Watch Company, Forbes Company, SA Girard-Perregaux Company, The Gruen Watch Company, Hamilton Watch Company, Tag Heuer Corporation, The Lejour Watch Company, Longines-Wittnauer Company, Omega Watch Company, the American Rolex Company, Seiko Watch Company, and Zodiac Watch Company.

Both Bulova and Omega planned to follow the "Buy American Act" requirements, meaning they had to ensure at least 51 percent of the components were either assembled or manufactured in the U.S.A. To comply with this, Omega manufactured their stainless steel cases for Speedmaster Professional in Luddington, Michigan, using Starr Watch Case Company.

The crystals were shipped from Switzerland to the Starr Watch Case Company (which no longer exists), where they were installed. The finished case with crystals was then sent to Hamilton Watch Company in Lancaster, Pennsylvania, for inspection and testing.

Due to the difference in gravity, the astronauts weighed only one-sixth of what they weighed on earth.

The case with crystals was transported to Switzerland for the works to be installed, and for final control and environmental testing.

At the same time, the Bulova Watch Company delivered sixteen chronographs to NASA for testing. It was later discovered that these watches had been manufactured in Switzerland, and that Bulova had purchased them through their daughter company in Switzerland, Universal Genève. The sixteen chronographs were dismantled by Bulova in their research lab and each was fitted with new crystals, new case, crown, bezel, hands, packs, and rings. The original works and back of each watch were retained.

Omega Speedmaster was re-certified in 1972, and in September 1978, NASA arranged for a number of new tests in order to certify an official wristwatch for use on future space missions. Omega Speedmaster was again the only watch to pass all tests.

Omega's mechanical Speedmaster is most likely history's most tested wristwatch—as the only watch to be qualified by NASA for all manned space missions, and used during Mercury, Gemini, Apollo, Skylab, Apollo-Soyuz, and the Space Shuttle mission. Perhaps the greatest heritage from Speedmaster Professional is its passing the stringent tests. Even now, more than 40 years after it was first introduced, it is still available and is still the only watch ever to be flight-qualified by NASA for extravehicular space activity.

If you haven't watched Ron Howard's movie *Apollo 13* starring Tom Hanks, you should. It is a must see. Omega Speedmasters are clearly visible attached to the space suits with Velcro, exactly as the real astronauts wore them. The astronauts also had a personal Speedmaster with a stainless steel strap, for when they were out of their space suits.

IN SEPTEMBER 1978 NASA ARRANGED A NUMBER OF NEW TESTS IN ORDER TO CERTIFY AN OFFICIAL WRISTWATCH FOR USE ON FUTURE SPACE MISSIONS. OMEGA SPEEDMASTER WAS AGAIN THE ONLY WATCH TO PASS ALL TESTS.

Omega's president Stephen Urquhart and Buzz Aldrin celebrate the 40th anniversary of the moon landing.

According to the NASA Explorers program, there are currently four wristwatches certified to fly in space. They are standard models available in stores. The certification ensures the watch can endure vacuum, extreme temperatures, and be able to expand without breakage or cracking. The first one is of course the Omega Speedmaster Professional. What a lot of people may not know is that this watch traveled with Walter Schirra on his way around the earth six times in the Sigma 7 spacecraft in 1962. This was before NASA officially certified Omega.

The second certified wristwatch is the Omega X33 Speedmaster Titanium Chronograph. This is a very sophisticated multifunction quartz

Astronauts Neil Armstrong, Michael Collins, and Edwin "Buzz" Aldrin were selected for the Apollo 11 mission.

watch and an extension of the original Omega Speedmaster. The watch has both analog and digital features. Night legibility is facilitated by strong luminescent material. The watch is durable and lightweight because of its titanium construction. The new Speedmaster has several timing functions for use during experiments or other closely monitored activities. The Casio G-Shock is another wristwatch certified for use by NASA's astronauts, and is useful when working on projects requiring precise measuring. The fourth certified wristwatch is the Timex Ironman. This pioneering watch has an LED port to synchronize up to 10 alarms to a calendar on a computer. It stores 38 phone numbers, identifies messages, and gives you the time in two different time zones.

Today, all watches used by NASA are the property of the state and must be turned in when the astronauts return to earth. Astronauts may check out the watches and bring them home to get to know them.

Both versions of the Omega Speedmaster, the Casio, and the Timex are all available for anyone to buy. Omega Speedmaster Professional also gives you the possibility to own a part of history.

ASTRONAUTS WHO WORE THE WATCH

Apollo 8	—————	Bill Anders
Apollo 8	—————	Jim Lovell
Apollo 10	—————	Tom Stafford
Apollo 11	—————	Neil Armstrong
Apollo 11	—————	Mike Collins
Apollo 11	—————	Edwin Aldrin
Apollo 12	—————	Dick Gordon
Apollo 13	—————	Fred Haise
Apollo 14	—————	Alan Shepard
Apollo 14	—————	Ed Mitchell
Apollo 15	—————	Al Worden
Apollo 15	—————	Jim Irwin
Apollo 17	—————	Ron Evans

MILESTONES

1957
Soviet launches its first satellite, Sputnik 1.

1957
The dog Laika is the first animal in space.

1957
Omega Speedmaster Professional is introduced. Later called "The Moon Watch."

1959
A soviet spaceship named Luna 2 performs the first crash landing on the moon.

1962
John Glenn is the first American into orbit in Mercury Atlas 6.

1963
Valentina Teresjkova is the first woman in space on board Vostok 6.

1965
NASA starts testing to find the official space program watch.

1966
Luna 9 lands on the moon.

1967
The astronauts Virgil Grissom, Edward White, and Roger Chaffee all die during an exercise when there is a fire in Apollo 1.

1968
NASA launches Apollo 8, the first manned spaceship to pass the moon.

USA's Mercury program
1961–1963

USA's Gemini program
1965–1966

USA's Apollo program
1967–1972

1973
The space station SkyLab
is installed.

1981
Columbia STS-1 is the first flying
vehicle in orbit.

1971
Saljut 1, the world's first
space station,
is installed.

1972
The last moon
landing with
Apollo 17.

1976
The Viking 1 probe
lands on Mars.

1969
Neil Armstrong and Buzz Aldrin land
on the moon.

1972
Omega Speedmaster Professional is
re-certified for use by NASA and the
space program.

1979
Europe enters the space age by
sending up the rocket Arian 1.

1998
Launch of the first part of the
international space station, with
16 different countries involved.

The US Space Program
1981–1998

FRANCK MULLER

For some a watch is a timekeeping device, for checking today's date or to make sure you reach your next meeting. Franck Muller's new Aeternitas Mega 4 is the world's most complicated timepiece.It has replaced the Patek Philippe Caliber 89, which used to hold that record. In watchmaking terms, complications mean, simply, additional functions beyond the regular time functions; for example, calendars, alarms, or moon phases. Aeternitas Mega 4 has no less than 36 complications, 1,483 different components, and 99 jewels.

I know you must be seriously into watches to understand all of this, but when you reflect on those numbers and the fact that 1,483 small parts are being assembled into a small case, surely anyone must find this impressive. It is more complicated assembling a small watch than it is to have a Rolls Royce handmade—hence the price

of 2.27 million dollars for the watch it took all of five years to design. That is, if Mr. Muller allows you to buy it... Franck Muller will send you a list of all 36 complications, and their website allows you to see them all as well. It has rightly been named "King of Complications." The first watch was submitted to its owner in a solemn ceremony at the Hotel de Paris in Monte Carlo.

The lucky owner of this prestigious timepiece was the collector Michael J. Gould, arriving in his private jet from his home in Colorado. Together with his family and friends, he was presented with the first Aeternitas Mega 4 personally by Franck Muller and Vartan Sirmakes. The presentation happened in front of 400 guests and representatives from the Monte Carlo authorities. This is also one of the most expensive watches sold outside of auction.

TIME IN WORDS

You can choose your text in German, English, or French.

Germany's Biegert & Funk offers in its products the principle of the time display in words. Their wristwatches include the QLOCKTWO W. At first, the display looks like a simple matrix of seemingly randomly placed letters, but with a push of a button, it shows the time in text, for instance, "It is seven thirty." Time is presented in five-minute intervals, while four small dots give the exact minute display on the screen. The company has won several awards for its special design and innovative functions. Similar to the award-winning model QLOCKTWO Classic, QLOCKTWO W also has a classic look.

It comes in high quality brushed stainless steel in 35 × 35 mm. The watch also shows the date, and is available with a 24 mm leather strap. You can choose text in German, English or French.

PILOT
WATCHES

PILOT AND DIVING WATCHES ARE THE TWO MOST POPULAR CATEGORIES WITHIN SPORTS WATCHES. PILOT WATCHES ARE ALSO CALLED AVIATION WATCHES, OR FLIEGER UHR IN GERMAN, FOR INSTANCE.

An American fighter pilot is getting ready for another mission wearing a Breitling.

One would think that pilot watches are reserved for pilots, but it is not that simple. Even though most pilots probably wear watches, and a lot of them wear a pilot watch, it has been a while since the watches had a real usefulness as a tool to navigate, measuring average speed and fuel consumption. Prior to the time of the calculator, some watches were equipped with a circular slide rule. This could convert from metric to standard measurements, and do speed, fuel consumption, and distance calculations. But it was relatively difficult to use and demanded both experience and practice.

A growing need emerged for watches showing several time zones. There is a world of difference of course between the needs of a pilot flying an Airbus 280 versus a pilot flying a small plane. At the time when the watch had a real function for pilots—before modern instruments and computers took over—the time in the air was also of high importance. Furthermore, the watch was an instrument for navigation and calculating fuel consumption in relation to flying time. This was mainly used by fighter pilots in the air forces and less for commercial aviation. In the 1930s, some watchmakers developed watches for different air forces. Producers like Hanhart, Junghans, A. Lange & Söhne, Tutima, IWC, Hamilton, Sinn, and Omega made watches for RAF, Luftwaffe, and the American Air Force during the Second World War.

If we go back in time and look at models used and purchased by pilots and flight crew, the most popular were Longines Lindberg Hour Angel, Omega Speedmaster, Omega Flightmaster, Rolex GMT, Glysine Airman, Fortis Flieger Chronograph, Breitling Navitimer, Breitling Cosmonaut, and Breitling Chronomat. Breitling and Rolex GMT were particularly hot because they were often used by pilots in commercial aviation.

If I were to recommend a good and functional pilot watch today, my choice would be the Omega X-33 and Seamaster 120 Multifunction in addition to Breitling's Aerospace and Emergency watch.

The Fortis Flieger Chronograph is a classic pilot watch.

One of my favorites among pilot watches is the IWC Big Pilot. IWC also recently launched Pilot Chronograph Top Gun Miramar. Miramar, California, is where TOPGUN pilots are trained in the U.S. The watch comes in titanium and a ceramic material and the case measures 48 mm. It has a power reserve of 168 hours, a flyback chronograph, and date. It comes in a matte finish with clear visible numbers and hands. The watch also has an army band. I particularly like the slightly worn retro style of these watches, and the sand-colored hands and matte case make it rugged yet stylish.

Fortis is another watch I really like and have in my collection. This company is relatively unknown, even though they've existed for more than 100 years. They have, for the most part, been making watches for the space and aviation industries, including their Cosmonaut and Flieger. The most classic and best-known model is Fortis Flieger Chronograph—a clean and stylish watch in a 40-mm sandblasted case, with double anti-reflective coated glass, fluorescent hands, date, and chronograph function.

Bell&Ross is quite a young company, established in 1992, and it specializes in watches with a look similar to airplane instruments. They have models with square cases formed like altimeters, turn coordinators, and horizon instruments. Rough-looking watches, but a little too big and quirky for my taste.

In the category of pilot watches, Breitling should not be overlooked, as it has always had a very clear aviation profile. This company has a long history of developing pilot watches with varied functions, views, and tables. The famous ones are Navitimer and Chronomat.

There are also stories about the origins of different models that were influenced by aviation. Rolex GMT Master was based on an idea by Juan Trippe, one of the top leaders of Pan Am, who worked with Rolex and developed

Bell&Ross is obviously inspired by cockpit instruments.

One of my favorites, IWC Big Pilot.

a model with date, four hands, a 24-hour view, and a bezel you could rotate. This watch was launched in 1954. All pilots at Pan Am received this watch; the entire production run was made for them. This increased the demand among other pilots, and eventually it was made available on the market.

The "flyback" function was originally meant for aerial acrobatics and precise maneuvers. Today, the pilot receives all his information from the control towers, computers, and instruments, so all the watch does now is mainly tell time.

Some elementary things, though, still apply to a watch in the pilot category. The size of the face, hands, and numbers is important for readability. At the same time, it needs to be light and avoid getting stuck. It should show analog time and have a large second hand, and in addition, it must be precise, waterproof, and shock resistant. It also needs contrast and anti-reflection coated glass for good reading visibility in bright sunlight or darkness.

Many brands would have liked to be associated with pilot watches, but their models are often not appropriate because they lack too many important functions. Despite this, the manufacturers use names on their models like Navitimer, Airman, Flieger, B-52, Skyhawk, Aerospace, Big Pilot, Top Gun Miramar, Spitfire, Linberg Hour Angle, and Pilot.

Some manufacturers use "Top Gun" effects and pilot helmets and suits in their advertising—clearly referring to our childhood dream and aiming to sell us men the cool pilot image!

PORSCHE DESIGN INDICATOR

This watch was originally made for the Porsche Carrera GT. A super car of the 2000s. A totally unique car, only 1,270 made, launched in 2004. It is a 5.7 liter mid-engine V-10 with 612 horsepower. The watch consists of approximately 800 individual parts and is assembled by 12 engineers and watchmakers, directly inspired by the car. The buttons on the side of the case are shaped like the pedals, and the back of the case, a sapphire crystal base, lets you admire the winding crown's splendid rotor in the form of a Carrera GT wheel rim. Porsche Design Indicator P'6910 is a world first in mechanical-digital stopwatch function display. It can record up to 9 hours and 59 minutes. The watch was launched after the last production run of the Porsche Carrera GT. It is available in three design variants. The photo shows a variant in rose gold with PVD-coated titanium.

PRICE: $225,000.

HUBLOT
ONE MILLION $
BLACK CAVIAR

Hublot is known for their special and flashy timepieces. Their models come in zebra and tiger patterns, carbon, and a whole range of different luxury materials. I truly like many of their models, but some are a little too much for my taste. I must admit I am fascinated by their wish to challenge this traditional industry. There are plenty of boring watches on the market. Hublot is Ferrari's official partner and has made a model series with the Italian car manufacturer. This is a fairly new cooperation begun when Hublot took over after Panerai.

The model series Big Bang now has a new edition: Hublot One Million $ Black Caviar. And yes, you guessed it. The price is one million dollars. It is obviously an extension of the far more attainable Big Bang Black Caviar—in ceramic. These are timepieces for people who crave attention, though Hublot claims they have a subtle look.

The Hublot One Million $ Black Caviar is festooned with an 18-karat white gold clasp and with hundreds of precision-cut black diamonds, a total of 34.5 carats. There are 322 diamonds on the case, 179 on the bezel and 30 on the clasp. Hublot made only one, so this is a rare piece! There are more than 2,000 hours of painstaking design and production work in this timepiece. It requires some exceptional skills to be able to baguette-cut diamonds this way and assemble them in a watch. This watch won the Grand Prix d'Horlogerie de Genève in 2009.

It builds on Hublot's design philosophy of "invisible visibility," designed to be subtle but at the same time so dramatically different you cannot help but stare. It has a diameter of 41 mm and is equipped with an HUB1112 automatic mechanical movement and a 42-hour power reserve, and is waterproof to 100 meters.

PRICE: $1,000,000.

As if all of this were not enough, Hublot also launched Big Bang 5 Million. This is one of the most expensive watches you can buy, with its 1,282 sparkling diamonds. It took 14 months to finish and was sold to The Hour Glass of Singapore. You can see it exhibited there unless it already has found a new home in the house of a passionate watch collector with a particularly fat bank account.

BREITLING EMERGENCY

**THIS WAS ONE OF THE FIRST WATCHES
I PUT ON MY WISH LIST.**

Big, powerful, and with a lot of exciting functions. Stories circulated about having to sign a contract with the civil aviation authority promising not to abuse the transmitter in this watch, and about skiers saving their lives with its ingenious technology. I guess this must be the only watch that has to be sent to the Breitling factory for service. None of Breitling's service centers have permission to open it.

The first version of Breitling Emergency was launched in 1988 with a patent. It had a pretty traditional milled case, but stood out because the hinges were two of the main components of the watch: the chronograph and the antenna. The antenna was wound around a capsule containing the micro-transmitter. Breitling put a lot of effort into increasing the battery capacity, and improving the range and user-friendliness.

The optimized and finished version of Breitling Emergency was completed in 1995, with another patent linked to the smart antenna system in the watch.

The watch has been thoroughly tested and it has saved lives. In 1977 it was used and tested during

a large exercise in Hong Kong performed by the Royal Air Force, U.S. Navy, U.S. Coast Guard, and the Japan Maritime Self-Defense Force. Also in 1977, it saved the lives of thirteen crew members of the *Mata-Rangi* when the ship was hit by a powerful storm off the coast of Chile.

The watch has gained cult status within aviation groups, is used by students and instructors in the U.S. Navy's TOPGUN program, and has also been used by the Blue Angels, the Thunderbirds, and Patrouille de France. These are demonstration and precision flying teams.

The two main functions are the electronic chronograph with all the necessary functions for

Different types of the model. Breitling has recently launched their new Emergency watches.

aviation, and the microtransmitter fixed at 121.5 MHz / 406.040 MHz, the international frequency for emergency calls. The watch has 48 hours of independent battery life.

To activate the emergency beacon transmitter, you manually deploy the antenna on the lower right side of the case and extend it to its full length. The signal varies somewhat, depending on terrain and area, and also the rescue aircraft's altitude. On top of a mountain, the signal carries 400 km to an aircraft flying at an altitude of 10,000 meters. On flat terrain or inside a boat, the length can vary between 36 and 160 km with a plane flying between 900 and 6,000 meters.

TAILOR-MAKE YOUR OWN WATCH

SPORTS STARS AND CELEBRITIES HAVE FOR A LONG TIME NOW HAD SPECIAL EDITIONS MADE OF THEIR FAVORITE WRISTWATCHES. RAPPERS HAVE HAD THEIR WATCHES SPRINKLED WITH DIAMONDS OR MADE IN LEOPARD PATTERNS. THE SULTAN OF OMAN HAD A ROLEX WATCH TAILOR-MADE IN 18-KARAT GOLD. THE BEZEL WAS MADE WITH 300 DIAMONDS AND SEVERAL RUBIES. HE REFUSED TO SELL IT, EVEN AFTER AN OFFER OF OVER $450,000.

A Rolex model made in a quantity of one only, for a customer. Matte finish and unique bezel.

Some companies deliver famous models in black matte DLC coating, while others make their own versions of older models based on giving completely new watches a vintage look. Some manufacturers also make their own versions of classic and older models with new leather straps with a worn look.

I talked to the owner of Project X Designs, Daniel Bourn, about what he thinks of customizing famous watches. He feels that a lot of the watches today are shiny jewelry and

models that everyone is wearing, and thus have lost some of their uniqueness; it's a loss of the simple, stylish, and functional.

His company offers special editions, made in limited numbers, of watches that were originally made for diving and military purposes. They are for those who yearn for something special. Project X Designs has made watches for celebrities like Mark Wahlberg, 50 Cent, Daniel Craig, Mark Webber, and Sebastian Vettel.

Project X Limited Edition Collection reflects the unmistakable inspiration of the early 1960s and '70s classic watches. It is based on historical models and the watch that was used by English Special Forces. (This is the watch Sean Connery probably would have worn in his early Bond days.) A matte bezel and G10 NATO "military band" gives the watch a clean and simple style. In my opinion, it is a watch for Special Forces members or divers. The watches in this series are based on the Rolex Submariner and are available in a limited edition of twenty-four and twenty-eight pieces.

Project X Design's Stealth – MKIII based on Rolex Submariner.

The case has "Stealth" in military green letters, and a logo for military special units for divers and paratroopers. Some may have noticed that the case is modified: the two "ears" protecting the crown on the original pieces have been removed. A small detail that I find enhances the look.

The company has several models and series based on Rolex Explorer, Submariner, Daytona, GMT Master, and Milgauss. Audemars Piguet's Royal Oak and Patek Philippe Nautilus have also received their special treatment.

Brevet + Primero Rolex Daytona in stainless steel.
Based on Zenith's legendary Striking 10th.

Brevet also makes its own special models, and I have especially noticed Brevet + Primero Rolex Daytona. The reason for this is my fascination with and interest in Zenith El Primero Striking 10th. That model was made in a limited edition of 1,969 and in my opinion is one of the finest timepieces ever made. It is with mixed emotions I see this watch with Zenith's unmistakable bezel with the three small "eyes" in gray, black, and blue with the fire-truck-red second hand. Fortunately, it is missing the Zenith star. I am not comfortable with this edition but I am still impressed with the expert work put into it. The model is made in an edition of only 33 at a price of $18,900.

**Patek Philippe Nautilus
in matte black.**

PAUL NEWMAN
AND HIS DAYTONA

STEVE MCQUEEN IS FAMOUSLY KNOWN AS "KING OF COOL," BUT IT IS DEBATED THAT THE OTHER ULTIMATE COOL GUY WAS PAUL NEWMAN.

"Paul Newman was the ultimate cool guy that men wanted to be and women admired," Arnold Schwarzenegger stated after Paul Newman passed away. "He was an American icon, a splendid actor, a renaissance man, and a very generous, but shy, philanthropist. He entertained millions of people in some of Hollywood's most memorable roles, and devoted his life to other people, especially seriously ill children. Paul Newman was a unique, one-of-a-kind man. The beloved film star will be missed by a whole world of fans and admirers."

Newman was up to a lot of things in addition to being an actor. In the 1980s, he opened a camp for critically ill children. He named it Hole in the Wall Gang from the movie where Newman played Butch Cassidy and Robert Redford was The Sundance Kid. There are nine Hole in the Wall Gang Camps around the world.

Paul Newman was a passionate motorsport and race driver.

In 1985 he started making his own salad dressing, Newman's Own, an instant success. Paul and his wife, Joanne Woodward, decided to donate all income from this business to charity.

And so...about Paul Newman and wristwatches... He grew closely attached to Rolex Daytona, the strikingly beautiful, timeless and sporty wristwatch.

Paul Newman first wore his Rolex Daytona in 1972. At the time it was not popular at all; the "Exotic Dial Daytona" had been displayed in shop windows for years without selling, at a price of around $300. Paul Newman was spotted with the watch on his wrist and the interest in it took off. It sold out and had long waiting lists. Today this watch model is worth $150,000.

So how does a change like this happen overnight? It is hard to tell exactly. The answer is probably a combination of super marketing by Rolex, and Paul Newman wearing the watch at the right time and making it as iconic as himself.

It is also quite interesting that as soon as Rolex stopped its advertising of the Daytona in stainless steel, its sales took off completely. I actually don't think Rolex has done any ads for this Daytona since the end of the 1960s. It sells itself.

I am not sure if it is entirely true, but it is claimed that throughout the years, Rolex has met only about 80% of the demand in the market for this watch. An element that, I suppose, increases the demand.

Rolex owners and watch enthusiasts may wonder why older models of the Rolex Daytona are referred to as "The Paul Newman Daytona." There are several versions of stories about Paul and his watch. The truth is that "The Paul Newman Daytona" is more than one single watch. He was a great fan of this watch and owned at least five different models.

So where does the saying come from? Newman played in a movie called *Winning*, together with his wife Joanne Woodward and Robert Wagner. Newman plays a race car driver, and he became so inspired by this role that he later started his own career as a race car driver.

"Paul Newman was the ultimate cool guy that men wanted to be and women admired..."

Arnold Schwarzenegger

One of my Rolex favorites: Model 6263 from 1970. Daytona with black dial and white "eyes" or sub-dials.

Some people referred to the stainless steel watch he was wearing in the movie as a Rolex Daytona. That turned out to be wrong. There were also speculations that the term "The Paul Newman Daytona" was created by auction houses a little too eager to sell Rolex Daytona watches as previously owned by Newman.

ROLEX INSPIRATION

Paul Newman was an authentic Rolex Daytona fan. At the end of 1970, Rolex sponsored a book that was published about Paul Newman's career. The book was published with two different covers. The first one, on the French edition, shows a PR photo of Paul Newman from the movie *Winning*.

The other cover, also sponsored by Rolex, shows Paul Newman with a Rolex Daytona exotic dial in stainless steel with a fat strap. This is a Rolex Daytona model 6241 or 6239. The watch is a mechanical Daytona, and is probably the watch that created the expression "The Paul Newman Daytona."

Rolex is known for its strategic marketing skills, so the obvious question is whether it consciously created this mystique by tying Rolex Daytona and Paul Newman together? It was an ingenious match. It created a model that became one of the really great Rolex successes. Customers waited years for this model.

Rolex historian John Brozek asked Paul Newman once in an interview about his connection with Rolex. Newman answered he never had a Rolex connection.

What most people associate "The Paul Newman Daytona" with is the exotic dial version: the name reflecting the black outer ring on the edge of the bezel, matching the small hands in the chronograph.

Newman with the Cosmograph Daytona, model 6240 from 1965. White bezel with black sub-dials.

In the cover photo of a book about Newman, he is wearing what likely is the Rolex Daytona Reference 5241. This version has small red sub-seconds. There is also a version with white seconds. Many watch lovers, me included, find this the grandest Rolex Daytona ever made. The combination of red, white, and black with stainless steel makes this watch a beauty. The simple writing on the bezel makes it clean and classy. This model still has a large fan group, and vintage Rolex Daytona models are ranked high in Hollywood.

The first Paul Newman Rolex model is exciting for many reasons. In an interview, John Brozek asked Newman where and when he got his first Rolex Daytona. Newman answered that he received his first Daytona as a gift from his wife, Joanne Woodward, in 1972, the year he started his professional racing career.

Interestingly enough, Paul Newman wore this Daytona a lot, but in early 1990, it disappeared and was never seen again. Did he lose it? Or give it to someone?

Later, he was observed with newer models of the Daytona. During the World Series GP in Mexico City, he was wearing a Rolex Daytona with white bezel and Zenith's mechanical engine and a steel strap. This watch has been referred to as Rolex Reference 116520 with Rolex's own engine, but this turned out to be incorrect. The model with the Zenith movement is characterized by the 3 and 6 in the lower chronograph eye situated on the same line.

You were one of the really cool guys if you were wearing a watch Paul Newman had once owned.

The other Daytona model Newman was spotted wearing is Daytona Reference 6263, with black dial and black bezel in stainless steel. This watch has screw down protectors on the two chronograph buttons, unlike the one with the white dial.

Paul Newman was also very fond of the fat strap with his Rolexes. As he grew older, he started using the Rolex Oyster strap. He also replaced the older straps on his vintage Rolex Daytona with a new Rolex Oyster strap with an extra lock.

Paul Newman was obviously into watches and would often be seen wearing one of his Daytona models. He liked the ones with two switchable push buttons. The first models did not have this function. A Rolex Daytona with the "turn and lock" buttons would be waterproof as long as these buttons were not unscrewed underwater.

When Paul Newman was over eighty years old, but still an active race car driver, he started wearing a new Daytona in white gold with a black crocodile strap. He had a great personality, with equal passion for both Rolex and racing.

ROLEX MOVES FROM ENGLAND TO SWITZERLAND

Hans Wilsdorf

In 1919, Hans Wilsdorf left England and moved his company to Switzerland, where it was established as Rolex Watch Company. The name was later changed to Montres Rolex SA, and finally to Rolex SA. England had introduced high taxes on the import of luxury goods and on the export of gold and silver, components of the cases Rolex used, which resulted in very high manufacturing costs.

When his wife died in 1944, he established the Hans Wilsdorf Foundation and transferred all his shares to the foundation to ensure that a portion of the surplus was used for charity. The company is still owned by a private foundation and no shares are traded on any stock exchange.

In December 2008, the CEO, Patrick Heiniger, abruptly left the company after sixteen years "for personal reasons." Rolex has always officially denied having lost a billion Swiss francs invested with Bernard Madoff, the American asset manager who pleaded guilty to the worldwide Ponzi scam.

CONCORD C1 TOURBILLON GRAVITY

Tourbillon shows the time more accurately, but is not critical for a watch's precision. But, it is a beautiful item all about mechanics and technical magic. The Concord C1 Eternal Gravity is made with 227 baguette-cut diamonds on the case and a further 100 cut diamonds on the dial. There exists only one watch, and it was offered at a staggering price of 1,300,000 Swiss francs (nearly $1,350,000.) The C1 Eternal Gravity from Concord is a one-of-a-kind with the outboard-mounted vertical tourbillon, visible through a small picture window in the side of the case.

C1 Tourbillon Gravity is presented in a 48.5mm case and contains a Concord Caliber C100 with 84-hour power reserve.

CERTIFICATION

C.O.S.C.

Contrôle Officiel Suisse des Chronomètres is a certification of Swiss movements. Chronometer Certification ensures a deviation of less than −4 to +6 seconds per day. This certification is mostly for rather expensive watches and ensures high accuracy.

Glashütte Observatory

As C.O.S.C certification is for Swiss movements only, German watch manufacturers have created their own similar certification. The German dealer Wempe has been issuing certifications since 2006 from an old observatory with a view over the city Glashütte.

Gran Seiko Standard

This is Seiko's own version of C.O.S.C. It actually imposes even stricter requirements to accuracy with a maximum deviation of −3 to +5 seconds per day.

Fleurier Quality Foundation

This standard was created many years ago by "Fleurier based" companies Chopard, Parmigiani Fleurier, Bovet, and Vaucher. In addition to quality, precision, and aesthetics of the movement, this also ensures the watch is 100% Swiss developed and manufactured.

EIGHT PAST TEN

HAVE YOU NOTICED THAT ALL WATCHES SHOW THE SAME TIME IN COMMERCIALS? IN ORDER TO GIVE THE BEST VIEW OF A WATCH FACE AND ITS DATE FUNCTIONS, IT IS ALWAYS SHOWN AT EIGHT PAST TEN (OR 10:08, TO BE PERFECTLY ACCURATE). THIS ALSO GIVES A VISUAL BALANCE AND SYMMETRY.

Watches with chronograph or GMT have individual ways of showing the extra hand or the small eyes.

MINUTE REPEATER AND PATEK PHILIPPE SEAL

The Geneva Seal.

Patek Philippe is known for its traditions and respect for the art of watchmaking. The firm's attention to details is unique, and its advertising states "you never own a Patek Philippe, you simply keep it for the next generation."

All models with a minute repeater must go through a last test at Patek Philippe before being delivered to the customer. Ever since its start, not one watch has left the factory in Le Roche without passing the expert listening test of the director of Patek Philippe. The minute repeater models have a distinct tone that receives serious attention.

A minute repeater is a complication in a mechanical watch that audibly chimes the hours (and often minutes) at the press of a button. There are many types of repeater, from the simple repeater which strikes the number of hours and quarter hours, to the ones that use separate tones for hours, quarters, and minutes.

The repeater originates from before there was widespread electricity for reading the time in the dark, and is a feature now also used by the visually impaired.

The minute repeater is a chime with three different tones. Hours are typically struck in a low-pitched tone, the quarter hours are struck with two gongs, and the minutes are struck on a higher-pitched gong. For instance, if the time is 02:49, a minute repeater will strike two low tones indicating two hours, three sequences indicating forty-five minutes, and four high-pitched tones to indicate four minutes.

The time is heard due to a small hammer striking a gong with different tones. Small coiled wires, springs, snails, and wheels make the minute repeater the ultimate single complication in watch making.

Patek Philippe is a family-run business. You will find the CEO, Thierry Stern, in his office listening to the timepieces just like his father and grandfather did before him. Never compromise on quality.

CELEBRITIES AND THEIR WATCHES

Celebrity	Brand	Model
ORLANDO BLOOM	ROLEX	VINTAGE GMT MASTER
JERRY SEINFELD	BREITLING	NAVITIMER
LEONARDO DICAPRIO	JAEGER LECOULTRE	MASTER MINUTER REPEATER
BILL GATES	CASIO	
MATT DAMON	ROLEX	EXPLORER
THE DALAI LAMA	ROLEX	DATEJUST TWO-TONE
PAUL NEWMAN	ROLEX	DAYTONA
ERIC CLAPTON	ROLEX	MILGAUSS
DAVID BECKHAM	ROLEX	SEA-DWELLER
JAMES GANDOLFINI	KOBOLD	SOARWAY DIVER
QUENTIN TARANTINO	JAEGER LECOULTRE	REVERSO DUO
QUENTIN TARANTINO	IWC	BIG PILOT
ALEXANDER SKARSGÅRD	IWC	BIG PILOT
TERRENCE HOWARD	TAG HEUER	TWIN TIME CARRERA
JON BON JOVI	U-BOAT	FLIGHTDECK CA 18
USHER	ROLEX	MILGAUSS SPECIAL EDITION
USHER	AUDEMARS PIGUET	ROYAL OAK OFFSHORE
PRINCE WILLIAM	OMEGA	SEAMASTER
RAFAEL NADAL	RICKARD MILLE	RM028
JON HAMM (MAD MEN)	JAEGER LECOULTRE	REVERSO GRANDE 980
SEAL	RICKARD MILLE	RM08
WILL FARRELL	IWC	PORTOFINO
PATRICK DEMPSEY	AUDEMARS PIGUET	ROYAL OAK OFFSHORE
SHIA LABOEUF	IWC	PORTUGUESE PERPETUAL CALENDAR
JEFF BRIDGES	ROLEX	VINTAGE SUBMARINER
BRAD PITT	PATEK PHILIPPE	NAUTILUS
STEVEN GERRARD	AUDEMARS PIGUET	ROYAL OAK OFFSHORE ROSE GOLD
	ROLEX	DAYTONA

BONO	ROLEX	PRESIDENTIAL WHITE GOLD
MICHAEL BUBLÉ	ROLEX	PRESIDENTIAL TWO-TONE GOLD
JASON STATHAM	PANERAI	GMT LUMINOR LTD. EDITION
FLAVIO BRIATORE	AUDEMARS PIGUET	ROYAL OAK OFFSHORE ROSE GOLD
TOM HANKS	BELL&ROSS	VINTAGE 123
KOBE BRYANT	NUBEO	CERAMIC CHRONOGRAPH
GREG NORMAN	AUDEMARS PIGUET	ROYAL OAK OFFSHORE ROSE GOLD
JUDE LAW	IWC	BIG PILOT
RICKY GERVAIS	TAG HEUER	MONACO CHRONOGRAPH
BARACK OBAMA	JORG GRAY	6500 CHRONOGRAPH US SECRET SERVICE
JUSTIN BIEBER	CASIO	G-SHOCK
RYAN REYNOLDS	HAMILTON	OFFICER AUTOMATIC
GEORGE CLOONEY	OMEGA	SEAMASTER PROFESSIONAL
DANIEL CRAIG	ROLEX	DAYTONA
ARNOLD SCHWARZENEGGER	AUDEMARS PIGUET	ROYAL OAK OFFSHORE LTD. EDITION
SYLVESTER STALLONE	PANERAI	RADIOMIR EGIZIANO PAM 341
CHRISTIAN BALE	JAEGER LECOULTRE	REVERSO GRANDE DATE PLATINUM
CHARLIE SHEEN	PATEK PHILIPPE	5970 PERPETUAL CALENDAR
HUGH JACKMAN	GIRARD-PERREGAUX	VINTAGE 1945 XXL IN WHITE GOLD
JAVIER BARDEM	CHOPARD	L.U.C. QUATTRO REGULATOR
COLIN FIRTH	CHOPARD	L.U.C. XPS
BILL CLINTON	CARTIER	BALLON BLEU TOURBILLON
STEVE MCQUEEN	CARTIER	TANK ALLONGÉE
JOHN F. KENNEDY	CARTIER	GULL
NICOLAS SARKOZY	ROLEX	DAYTONA
DUKE ELLINGTON	PATEK PHILIPPE	1946 1563 CHRONOGRAPH
ALBERT EINSTEIN	LONGINES	1929 RECTANGULAR MODEL

WATCHES' ROLES IN MOVIES

I EXPECT YOU HAVE NOTICED THE WATCHES CARRIED BY THE MAIN CHARACTERS IN MOVIES. ANYTHING BUT A COINCIDENCE, OF COURSE. ALL WATCH MANUFACTURERS KNOW HOW IMPORTANT THIS FORM OF MARKETING IS.

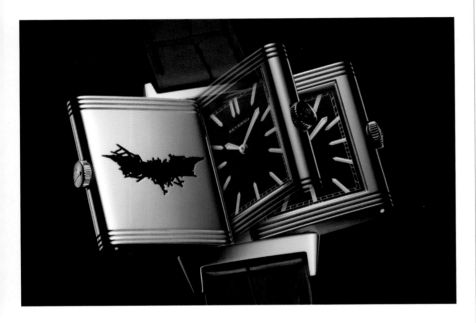

The Jaeger LeCoultre Grande Reverso Ultra Thin model that was launched in connection with the *Batman* movie.

Product placement is integrated firmly into global film production, and watches have become a very important part of it. Which wristwatch the main character is wearing is certainly not insignificant. It is noticed, and so should preferably be realistic and believable. As in real life, film watches are, for both men and women, a way to give signals about who you are, your tastes, and your style. The best product placement provides hints and signals about the character. The watch can also have an underlying meaning, and in that sense it endeavors to get the character to fit the watch. In the movie *Batman Begins*, Bruce Wayne wears the classic cult-watch Jaeger LeCoultre Reverso Grand Date. This watch can be flipped and has two sides, not unlike Batman's personality.

JAMES BOND AND HIS WATCHES

James Bond is, I guess, one of the most famous examples of the wide use of product placement. Men wish to be Bond, women wish to be with him, and some of us want his watches. From the start of the Bond movies, the main character has been linked to watches. He has shown Rolex, Breitling, Hamilton, Seiko, and Omega on the big screen. Omega Speedmaster has been used in several movies. Cars like Aston Martin and BMW also hold key roles in the movies.

The watches and cars have been assigned functions we can only dream about. There are advanced lasers, and remote controls for detonating bombs and for placing climbing hooks. Originally, Ian Fleming (who had his own steel Rolex Explorer) gave 007 a sturdy Rolex Oyster Perpetual. Since then, James Bond has, in going through different actors, been given several different Rolex models. In 1962's *Dr. No*, we got a glimpse of a Submariner on Sean Connery's arm. It was never described in the books whether the secret agent had a Rolex diving watch; the watch in the movie was used to describe James Bond's character and lifestyle. Sean Connery used this model throughout most of the Bond movies he acted in, and collectors

JAMES BOND WATCHES

Dr. No	Rolex
From Russia with Love	Rolex
Goldfinger	Rolex
Thunderball	Rolex, Breitling
You Only Live Twice	Rolex
Casino Royale	Rolex
On Her Majesty's Secret Service	Rolex, unknown
Diamonds Are Forever	Rolex
Live and Let Die	Hamilton, Rolex
Man with the Golden Gun	Rolex
The Spy Who Loved Me	Seiko
Moonraker	Seiko
For Your Eyes Only	Seiko
Octopussy	Seiko
A View to a Kill	Seiko
The Living Daylights	TAG Heuer, unknown
License to Kill	Rolex
GoldenEye	Omega
Tomorrow Never Dies	Omega
The World Is Not Enough	Omega
Die Another Day	Omega
Casino Royale	Omega, Longines
Quantum of Solace	Omega, Hamilton
Skyfall	Omega

are referring to it as the "James Bond Submariner." Performances like this often result in cult status for a watch.

TECHNIQUE AND GADGETS

Later, and in a very timely way, the agent received messages from MI6 on his digital watch. Received messages could be printed on tiny paper strips. The Seiko M354 was used in *Moonraker* with a built-in detonator for bombs, while in *Octopussy* one could spot a Seiko G757 Sport 100 with GPS tracking and a built-in satellite receiver. In the 1990s Pierce Brosnan took over the Bond role, with a new and slightly different edge. Rolex was substituted for Omega, which has a long history in the British military. Omega SeaMaster was the model the Royal Navy divers used in the 1960s. Ever since, Omega had been James Bond's wristwatch. Although the models did vary; in *Casino Royale* Daniel Craig wore a SeaMaster Planet Ocean. It fits his character, which is a little rougher than Brosnan's Bond. In *Skyfall* the agent wore both a Planet Ocean and a Seamaster Aqua Terra. There are discussions about the right Bond watch, whether it should be a Rolex or an Omega. Perhaps James, after a hard day of fighting and car racing, comes home and sets his Omega in the winder and puts on his vintage Rolex.

DON DRAPER

The reputable Swiss watchmaker Jaeger LeCoultre has also become aware of the popular series *Mad Men*, about the advertising industry in the 1950s. In the television series Jon Hamm, who plays the character Don Draper, is wearing a Jaeger LeCoultre. In the first season the watch was a Vintage JLC Memovox model and later a Reverso Classique. I was amused by this choice of watch—totally credible, and a perfect fit to the character. The company has produced a limited edition of 25 watches for hardcore fans: Grande Reverso Ultra Thin 1931 is a tribute to the series.

As a tribute to the TV series *Mad Men*, Jaeger LeCoultre has created a very limited edition series of 25 pieces: Grande Reverso Ultra Thin 1931 for hardcore fans. The advertising agency logo is embossed on the back of the case.

HISTORICAL REFERENCES

In some cases it is more important to find watches that represent the era the film is set in, than to match the watch with the main character's personality. In movies where the set is dating back in time, the credibility will falter by not using correct watches. *Pearl Harbor* had closeups of a Hamilton on Josh Hartnett's wrist. Hamilton delivered watches to the U.S. Army, Navy, and Marine Corps during the Second World War and is the correct choice for this movie.

In the 2003 movie *Master and Commander*, starring Russell Crowe, Swiss manufacturer Brequet was asked to reproduce a watch from the seventeenth century. Brequet accepted the challenge and made

Omega Speedmaster.

an exact reproduction of a stunningly beautiful pocket watch. More than 100 million people have seen Russell Crowe in this movie, carrying the watch, which perhaps was the reason Brequet never charged for this work. The famous Omega "Moon Watch" was used to time Apollo 13's journey into the Earth's atmosphere and helped bring the crew of the space shuttle safely back to Earth in 1970. In the movie *Apollo 13*, about the Americans' race to be the first on the moon, the Omega Speedmaster was of course visible in several scenes.

Hamilton is in many ways the natural choice for superstars and movies. The brand has been present in more than 300 movies and is known, for instance, from *Men in Black* and *2001: A Space Odyssey*.

Elvis Presley chose the American brand Hamilton and the model Ventura for the romantic movie *Blue Hawaii*. The quite spectacular shell-shaped watch is visible in many scenes, including the opening sequence where Elvis arrives in Hawaii after his military service. The watch was later sold at auction to the Hamilton historical collection.

TAG HEUER MONACO

In 1969 Tag Heuer launched the breakthrough watch with a square waterproof case. This made headlines, and one year later it was used by Steve McQueen in the famous race car movie *Le Mans*. The original watch had a black alligator strap and a blue dial, and featured a Chronomatic Caliber II movement. This made the watch a famous as well as a classic icon. The plan was for Steve McQueen to drive, and have the race legend Jackie Steward as his navigator. But the car, a No. 26 Porsche 917K, was not approved by the insurance company; neither were Steve McQueen or Jackie Steward. Forty years later, the watch known as the "Le Mans watch" is still in Tag Heuer's model series. New models have been launched, including the first watch driven by a belt drive mechanical advantage.

One watch used by Steve McQueen in the movie was expected to sell for several hundred thousand dollars at auction, but sold for only $7,990.

Tag Heuer launched several really nice versions of this model. To celebrate their forty-year anniversary in 2009, they launched a special version in an edition of 1,000 pieces, with blue dial and with Jach Heuer's signature, to honor their ambassador Steve McQueen. Another model has the Gulf logo on the dial to acknowledge their relationship.

There are many amusing examples of the wrong use of watches. In the movie *Braveheart*, they clearly forgot to tell the extras to remove their modern wristwatches before they got into battle.

PANERAI L'ASTRONOMO LUMINOR TOURBILLON 1950 EQUATION OF TIME

Panerai is from Florence, Italy. The company was founded by Giovanni Panerai in 1860, when he opened a shop, l'Orologeria Svizzera, near the Ponte Grazie bridge in Florence. It was the official supplier of watches for Regia Marina, the Italian Navy. The models have a case designed and manufactured by Panerai, but the movements are Swiss, by Rolex amongst others. Some models have become icons, like Radiomir. Panerai is very popular with celebrities and collectors. Sylvester Stallone spotted a Panerai in a shop window in Rome while filming *Daylight* and bought it immediately. Other Panerai owners are Jason Statham, Pierce Brosnan, Orlando Bloom, Bill Clinton, Dwayne Johnson, Ben Affleck, Heidi Klum, and Josh Harnett. Panerai has also cooperated with Ferrari to create a line of Ferrari watches. This deal has now been taken over by Hublot. This watch was presented in 2010 in the Salon International de la Haute Horlogerie in Geneva as a tribute to Galileo Galilei. L'Astronomo was made in a limited edition of only 30 pieces.

Upon ordering this watch you are asked to state where you live. This will be engraved on the back of the case next to a rotating star chart. The dial shows sunrise and sunset for the chosen city. Your city?

MONDAINE
& APPLE

The Swiss Federal Railway (SBB for short) is famous across the world for its punctuality. In almost all railway stations in Switzerland, hours, minutes and seconds are clearly visible for all travelers from the clocks hanging on the walls to ensure everyone that the trains are on schedule. Mondaine Watch Ltd. was established in 1951 by Erwind Bernheim. The company, now run by his sons, is perhaps best known for the iconic clocks that adorn the Swiss Railway stations. The famous clock design was launched as the Swiss Railway Watch in 1986. The characteristic red second hand glides smoothly but stops on 58.9 seconds, waiting for a radio signal to coordinate to the correct time. Mondaine's quartz wristwatches did not have this function, but complaints from customers made them relaunch a new and slightly more expensive version with the function included.

Apple used Mondaine's clock design in their operating system.

Mondaine holds the license for the watch design, and an employee by the name of Hans Hilfiker is the designer. The watches were trademarked in 1944 and have been honored by the Museum of Modern Art in New York and London Design Museum. This clock and its design have, through the decades, been the symbol of Swiss punctuality.

Apple announced iOS 6 in 2012, featuring exactly the same clock design used in its new operating system. Swiss authorities and SBB were proud of the fact that Apple chose their well-known symbol, of course not without a certain amount of compensation. Apple ended up paying $21 million to license the design.

PLOPROF —
OMEGA'S FIRST
DIVING WATCH

AN AD FOR THE ORIGINAL OMEGA PLOPROF STATED, "IT MAY NOT LOOK PRETTY ON THE SURFACE, BUT DEEP DOWN IT'S BEAUTIFUL."

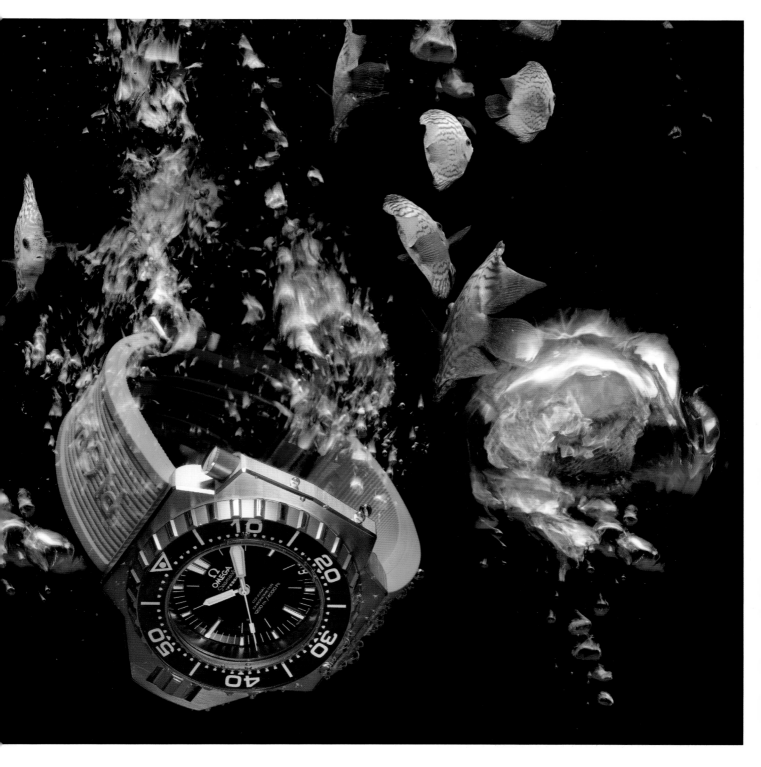

A strange statement for an official advertisement, but it nicely sums up the fact that Omega was one of the lead innovators, along with Rolex and Blancpain, in the market to supply serious instruments for professional divers (notably the likes of Jacques Cousteau). In its heyday, the Ploprof (a contraction of "Plongeur Professional") was an extremely advanced tool based on years of development.

Omega achieved a design that not only withstood the depths, but was able to remain underwater for long periods of time. It had several features that were unusual at the time. It was user-friendly, and had an easy-to-grip safety bezel and secure crown. It was arguably the best diving instrument of its time, but it was also very expensive. As a professional instrument, it actually sat at the top of the Omega product line, and was extremely expensive even for professional use. The Ploprof was a hit with men, but Omega made excuses for the look and it turned out to be not so popular with the ladies. The design isn't exactly elegant, with its fat orange minute hand and the strange-looking asymmetrical case. Even today, it resembles a Star Wars spaceship.

Still, we find the beauty in this wristwatch. So much beauty, in fact, that the original models are still enjoying their cult status and are sought after by collectors. A Seamaster 600 first generation Ploprof is worth about $30,000 in the market, and the second generation is worth about $15,000. The difference between the two can be spotted on the dial. The first generation watch

Omega Ploprof is mainly a diving watch, but has obtained a cult status among collectors.

has "600" engraved on the dial just underneath the Seamaster logo, while the second generation has an engraved "600/2000ft."

On the first model, the crown was situated on the right-hand side of the case, with the orange button situated on the left-hand side to be able to rotate the bezel. It was also called Ploprof Inversè. On all later models that has been mirrored. In 2009 Omega announced an updated

"It may not look pretty
on the surface, but deep
down it's beautiful."

Omega ad for Ploprof

version of Ploprof under its Seamaster line and could then offer fans of this design a very special model, a way to enjoy the famous diving watch.

Why do we find Ploprof so attractive? Hard to say, but I think it is for the same reasons we find well-developed tools and instruments attractive. Omega Proprof was born in a lab with engineers and technicians...no trendy designer enjoying his espresso in a café while discussing how one might enhance the design to make it attractive and exciting. It became an anti-luxurious high-end watch. The real thing it reflected was the result of cooperation and problem solving. It also represented innovation without worrying about a budget, and was designed to be a necessary tool when exploring the deep sea. Tools like this are made by smart people who inspire others to move forward.

The modern Omega Seamaster Ploprof 1200m still holds to these principles. It certainly gives that impression anyway. Mechanical diving watches are no longer exclusively for professional divers, but should you want to go diving, this watch is ready anytime. Held side by side with the original Ploprof, it looks alike. But it really is a modern reincarnation with important changes and upgrades. In a way, the new Ploprof is what Omega always wanted to build.

188

ed to let out helium fast enough. The two most extraordinary functions the watch has are the rotation system on the bezel and the protection of the crown.

Let's start with the tall, slightly-angled bezel; the insert is made of sapphire instead of acrylic like the original's was. Today most bezel inserts are made of ceramic, which is generally cheaper. The sapphire bezel insert has a standard minute marker with fully laminated numbers. The dial itself is superluminated. As opposed to most diving watch bezels, the Omega Proprof rotates in both directions. It makes it easier to set it...although the bezel itself cannot be moved without pressing the integrated orange button. The original model had this piece in plastic, but Omega installed an orange aluminium ring on the 1200m model. It could make it difficult to press and turn wearing diving gloves. The date is situated between 4 and 5 on the dial.

Another detail is the part that protects the crown. It has its positive and negative sides. On the one side, it protects the crown from pressure when descending, but on the other side, it is less comfortable to wear. If you remove the protection, the crown will function just like any other watch. Seamaster Ploprof 1200m has an updated, better-quality dial compared to the original, even if they do look very similar. The dial is clearer and easier to read, with a more elegant look, but with the same Omega Ploprof DNA.

While the original Ploprof was 54 mm wide and 45 mm tall, the new one measures 55 mm wide and 48 mm tall. This altering makes it look better and it fits better on the wrist. There is a lot of steel in this watch, so it is not exactly lightweight. Wearing it on your arm has been described as a workout. Notice the impressive details. The polished, sturdy, and well-integrated parts. The watch is waterproof down to 1,200 meters, twice what the original model offered. (Actually, the unofficial opinion is that both the original and the new model will handle even deeper dives than stated; not that I am telling you to try it out.) The case has an automatic helium release valve to ensure a deep dive will not cause helium to leak through the anti-glare treated sapphire glass. This is mainly for use in pressure chambers during ascent. The valve is construct-

The original on top, and the new model.

The watch was designed with professional divers in mind.

Omega Seamaster Ploprof comes in white, orange, or with a black rubber strap, but it is hard to overlook the shark-proof strap. While other similar straps can be uncomfortable, this is surprisingly comfortable on the wrist. The lock has two buttons. One is sturdy and great for diver extension, while the other is for adjusting size and fit.

Even after a few years on the market, the Seamaster Ploprof 1200m is a watch that sells well, although it doesn't come cheap. On the inside of every Omega Seamaster Ploprof 1200m there is a movement made by Omega. The Caliber 8500 automatic co-axial escape-ment with 60-hour power reserve is a great watch and one of Omega's first movements. The original Ploprof also contained an Omega-manufactured movement called Caliber 1002 automatic.

The relaunch of Seamaster Ploprof was a great success and to many collectors' and enthusiasts' liking. It was perhaps only natural that Omega presented a version of this iconic model at the Only Watch auction—ironically enough in 18-karat solid red gold. And just like Blancpain's Fifty Fathoms, the bezel is made in crystal for protection.

"If you think it's expensive to hire
a professional to do the job, wait until
you hire an amateur."

Red Adair

RED ADAIR

With his special name, red overalls, and a belt filled with gadgets he designed himself, Red Adair is a living legend and superhero. We in Norway remember him well from the biggest oil well blowout ever in the North Sea, on the Ekofisk 2/4 B Platform, on April 22, 1977. Asger "Boots" Hansen and Red Adair managed to stop the blowout with a bucket.

The guy with the ginger hair and great personality managed to promote himself well. He wore red shoes, red clothes, and drove red cars. All Red Adair Co. Inc equipment was painted bright red.

In 1962, he received a global award for extinguishing a fire in Algeria. The fire in the gas field was named "The Devil's Cigarette Lighter" after burning continuously more than five months with a 137-meter high column of flame.

Red Adair in a familiar pose, wearing a Rolex.

Hiring Red Adair Co. was expensive and the main man himself expressed boldly: "If you think it's expensive to hire a professional to do the job, wait until you hire an amateur."

Paul Neal "Red" Adair was born in Houston, Texas. He is said to have given everyone he worked with at Ekofisk a Rolex.

For Red Adair was also known for his connection with Rolex. He faithfully used his gold Rolex Oyster Perpetual, which became one of his many features. Rolex loved the celebrity and frequently put him in their ads: "If you ever need proof that a Rolex needs no pampering, watch Red Adair in action."

Red Adair Co. Inc was founded in 1959, and during Adair's long career he fought spectacular fires and oil catastrophes all over the globe.

SPACE INVADERS

ARE YOU ONE OF THOSE PEOPLE WHO PLAYED SPACE INVADERS AT A YOUNGER AGE? THIS POPULAR GAME TRIGGERED A CRAZE AND A SHORTAGE OF 100-YEN COINS WHEN IT WAS FIRST LAUNCHED IN JAPAN IN 1978.

Space Invaders and its pixel creatures were the very first iconic form of entertainment. Children and young adults spent thousands of hours in front of the screen pictures of this legendary cult symbol. Absolutely 1980s pop culture at its best!

No matter what your age, you can now get hold of a Romain Jerome Space Invader timepiece. Well, at least in theory. It comes with a fat price tag. The company is renowned for their Swiss quality watches and have launched several very special pieces over the last couple of years.

This fun watch is made by Romain Jerome together with Taito Corporation, which holds the Space Invader rights. The watch has an analog

No matter what age you are, you can now get hold of a Romain Jerome Space Invader timepiece. Well, at least in theory. It comes with a fat price tag.

display of time and a Swiss Romain Jerome Caliber RJ001, a modified ETA 7750 movement with 42-hour power reserve. It has a black PVD-treated case measuring 46 mm in diameter. Original fragments from the Apollo 11 space capsule are mixed into the case. Each of the characters is put together by a steady hand and they are hand painted, one by one. The design on the watch is made to look like the landing vehicle. There are four elements surrounding the case, made to resemble the landing feet of the space capsule. The back cover of the case has been carved out from a Moon Silver JR alloy, consisting of silver and (impressive) moon dust. The dial features graphics from the game, and all figures glow in the dark.

The company has outlets all over the world and the Space Invader watches are available in red, green, purple, and yellow. They are made in a limited edition—eight pieces of each color, with an asking price of around $19,000.

Romain Jerome Titanic DNA – Day & Night Tourbillon with
manual-wind movement, 46 mm case with deep black coal dial,
obtained with coal recovered from the Titanic.

A beautiful DeLorean with the characteristic "seagull wings" is also known from the *Back to the Future* movies.

Even if Space Invader fans want it, it's unfortunately not possible to play the actual game on the watch.

The Swiss watchmaker is known for his extraordinary watches. From moon dust in the case to watches with parts from the *Titanic*, Pac Man, and Space Invaders. A while back, Romain Jerome revealed an exciting project with DeLorean: a DNA theme where they used parts of metal from the sports car DeLorean DMC-12. This iconic car became well-known from the *Back to the Future* movies starting in 1981, and it has therefore been made in a limited edition of 81 pieces.

The Romain Jerome DeLorean DNA Watch can be yours for a mere $15,900.

Manuel Emch, director of Romain Jerome (and a Space Invaders fan), started his career in 2001, when Swatch Group gave him the responsibility for starting up the brand Jazuet Droz. Manuel Emch spent ten years in the company, working on building its philosophy, identity, dealer network, and plans for a totally new production line. However, he did not agree with the company's strategy and finally moved on.

The majority of Romain Jerome's models have prices ranging between $9 and an incredible $250,000 for the more extravagant tourbillons.

BREITLING NAVITIMER

Breitling Navitimer became a hit among pilots quickly after its launch in 1952. This classic and well-known watch has its own appeal to aircraft enthusiasts and has been launched in a great number of versions. This icon is Breitling's most known and popular model. It is still a classic and a little retro in its design and easy to recognize. Since the famous and classic 806 model from 1952, Breitling has managed to maintain the characteristics of the world's oldest chronograph. It has always been marketed with a link to aviation, and has John Travolta as its ambassador.

In 1992, Breitling launched a new version of Navitimer which at the time was the world's smallest mechanical chronograph. The model was 38 mm and was designed to fit any wrist. Since then, Breitling has made Navitimer models like Heritage with Breitling caliber 35 and World with Breitling caliber 24. The models vary in diameter from 38 to 46 mm.

Navitimer Airborne came in 1993, with a chronograph version limited to measure up to 3 hours and 10 minutes for more accuracy. Breitling launched the 1994 model with "rattrapante" and perpetual calendar. Rattrapante means that the chronograph function has two second hands on top of each other. The user can stop, start, and reset one with the other still going. This is also called "split-second chronograph" or simply "split chronograph."

Not without reason has this been the watch for pilots. It has a circular slide rule for different calculations. It can calculate ascent and descent rates, conversion of nautical miles to kilometers, etc. Another table can calculate the average speed, elapsed distance, and fuel consumption.

The first Navitimer model was the official selectionof AOPA (Aircraft Owners and Pilots Association) to follow pilots worldwide.

The Breitling Navitimer and Emergency are my personal favorites.

Breitling Navitimer 01 – Chronograph, 70-hour power reserve, black leather strap, and a diameter of 43 mm.

TUDOR – IN THE SHADOW OF ROLEX

DO YOU LIKE ROLEX BUT FIND IT TOO ORDINARY, OR PERHAPS IT ISN'T QUITE YOU? TUDOR IS A REALLY SUPERB ALTERNATIVE.

1970

2010

Tudor has relaunched several sporty and exciting models in the "Heritage" series: new watches based on tradition and history.

Tudor may seem a cheaper alternative to Rolex, but for watch enthusiasts, these watches are among the best in their class. High quality, a familiar name for watch lovers, a cool profile, and a totally independent brand with its own models. If you are looking for a different watch and don't want the Rolex stamp on your wrist, Tudor could be your alternative.

Of course, there is little doubt that Tudor has benefited from its connection to Rolex, and the brand is still selling through registered Rolex dealers.

Back in 1905, Hans Wilsdorf started his own company in England, manufacturing watches of high quality. He registered Rolex as a brand in La Chaux-de-Fonds in Switzerland in 1908, but the company stayed in England until 1920 before moving on to Geneva.

Hans Wilsdorf had a vision of creating a new brand that would be more easily obtainable for watch enthusiasts, but still have the same reliability and value that Rolex had become known for. Still, it was not until 1946 that Wilsdorf first launched Tudor. The name was chosen because he wanted to honor the Tudor period in England. Rolex had, before the launch of Tudor, offered a number of cheaper watches with names now forgotten: Marconi, Unicorn, and Rolco—watches they believed would increase sales.

Tudor Heritage Chrono Blue. Stainless steel case. 42 mm. Automatic movement caliber 2892 with chronograph function. Water resistant to 150 meters, 42-hour power reserve, sapphire crystal and with rotatable bezel.

Tudor Heritage Black Bay is a classic and stylish wristwatch.

Tudor has for years been relatively unfamiliar to people and has stayed in Rolex's shadow—despite the fact that this is a high-quality brand. Tudor was less expensive than Rolex and was unfortunately nicknamed "the poor man's Rolex." Perhaps not surprisingly, as the company made ads positioning Tudor as "the working man's watch." Tudor used the Tudor rose, the symbol from the Tudor dynasty, on the bezel.

The French navy (Marine Nationale) used Rolex Tudor watches for their divers. At the end of the 1960s, they bought the first Rolex Tudor Submariner. These are the same models used by the American Navy's elite divers and Navy SEALs.

In 2010 Tudor introduced a watch they called the Heritage Chronograph, inspired by the early 1970s, and the model was launched in the U.S. in an attempt to get rid of the reputation of being a lesser Rolex edition.

The watches cost about two-thirds the price of an equivalent Rolex, but hold the same quality. This particular model is strikingly beautiful and has the right vintage look; its price is affordable, considering its functions and quality. The Heritage Chronograph meets all the playbook requirements for a commercial success, but theory is not always reality. Image is everything, and so for success, the buyers must understand and appreciate the philosophy. The market liked it, and so do serious Rolex vintage watch collectors. The Heritage Chronograph is already an icon and highly valued by watch collectors and enthusiasts.

Tudor's Big Block Chronograph is often jokingly referred to as "the smart man's Daytona."

PATEK PHILIPPE

ACKNOWLEDGED AS BEING ONE OF THE FINEST TIMEPIECES IN THE WORLD AND ONE OF THE INDUSTRY'S STRONGEST BRANDS, IN ADDITION TO HOLDING THE RECORD FOR THE HIGHEST PRICE EVER AT AUCTION.

The company is today known as Patek Philippe, established in Geneva in 1839 by the Polish nobleman Telle Antoine Norbert de Patek and his fellow countryman François Czapek. The first watches were signed "Pate, Czapek & Co" until 1845, when Czapek left the partnership. Years later, the French watchmaker Jean Adrien Philippe became a new partner. Through the decades, celebrities have been wearing Patek Philippe—presidents, movie stars, and queen Victoria of England are all names on the customer list.

As early as 1867, at the Paris Exhibition, Patek Philippe showed watches with functions that were to become standard for complicated

A Patek Philippe timepiece from 1928 in 18-karat white gold with oval case and "single button." The chronograph was sold for $3.6 million at an auction at Christie's in 2011.

timepieces in the beginning of the twentieth century: a perpetual calendar, a repeater, and a chronograph with split seconds.

The two most complicated watches of all time were made by Patek Philippe. The first, made for Henry Graves Jr. in New York, was finished at the beginning of the 1900s and the other, Patek Caliber 89, the world's most complicated watch, was finished in 1989 (hence the name) and marked the company's 150th anniversary.

After the world was divided into twenty-four time zones in 1870, watchmakers everywhere tried to develop a device that could show the time in at least two cities in the world at the same time. An ingenious mechanism was later created by the independent watchmaker Louis Cottier from Geneva. Cottier made several models of "Universal Time" for Patek Philippe.

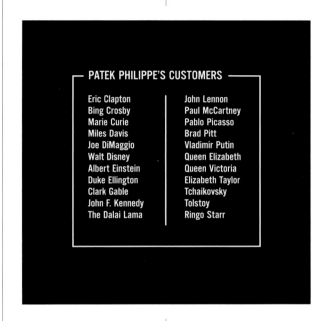

PATEK PHILIPPE'S CUSTOMERS

Eric Clapton	John Lennon
Bing Crosby	Paul McCartney
Marie Curie	Pablo Picasso
Miles Davis	Brad Pitt
Joe DiMaggio	Vladimir Putin
Walt Disney	Queen Elizabeth
Albert Einstein	Queen Victoria
Duke Ellington	Elizabeth Taylor
Clark Gable	Tchaikovsky
John F. Kennedy	Tolstoy
The Dalai Lama	Ringo Starr

Patek Philippe was also the first of the Swiss watch manufacturers that really managed to exploit its business in the United States. It signed an exclusive agreement with Tiffany & Company in New York—a major breakthrough. The economic crisis in 1929 meant production was reduced during the 1930s, but even while the economic problems continued, the company still developed new and innovative models. The most complicated was the famous Calatrava, with its perpetual calendar, minute repeater, moon phase, and triple date.

The name has historical roots in the Middle Ages, when a Spanish religious order defended the Calatrava fortress against the Moors. At the end of the nineteenth century, Patek Philippe decided that the emblem of one of the brave Spanish knights was to be its symbol, and this cross still marks all the watches.

Calatrava is still in the company's collection. In 1976 Patek Philippe introduced its sports model Nautilus, and in 1993 came the Gondolo series. The beautiful Patek Twenty 4, which in my opinion is the most beautiful women's model, launched in 1999 and was a modern interpretation of the Gondolo. Other models are the Calendar, Geneva, Grand Complications, Maxi Marine, Tourbillon, and Golden Ellipse.

In 1932 Patek Philippe was taken over by Charles and Jean Stern. Today it is run by the third generation in this family: president Henry Stern, and his son, vice president Philippe Stern.

PATEK PHILIPPE 324 S QA & THE GENEVA SEAL

Patek Philippe's iconic 324 S QA LU 24H was designed and developed in 1992 and is one of the company's most sought after movements due to its user-friendly perpetual calendar. A complication like this is normally difficult to use.

The calendar shows date, day, and moon phase and needs to be set once a year on the date March 1. It is put together by hand and is of particularly high quality with the coveted Geneva Seal, or in French, Poinçon de Geneva.

The beautiful Nautilus model from Patek Philippe in gold and steel. Moon phase and with brown crocodile leather strap.

The Geneva Seal is a mark of quality issued by the city of Geneva—a certification for wrist-watches and pocket watches, created in 1886. Only approved and inspected watches carry this seal. The seal attests to the quality and finishing of watches, not precision chronograph function or accuracy of time shown.

The seal clarifies that the watch possesses all necessary properties required for the award. Although it is mainly for the finish and decoration

of the movement, it is considered a high ranking in the industry. These Geneva watchmakers regularly submit their movements for the Geneva Seal certification: Patek Philippe, Cartier, Chopard, Roger Dubuis, and Vacheron Constantin.

The certification requires at a minimum that the watch has been made by qualified Geneva craftsmen from the city, or at least from a canton of Geneva. A similar certification is Qualité Fleurier (Fleurier Quality), reflecting precision testing. Participants in this certification process include Bovet Fleurier, Chopard, Parmigiani Fleurier and Vaucher.

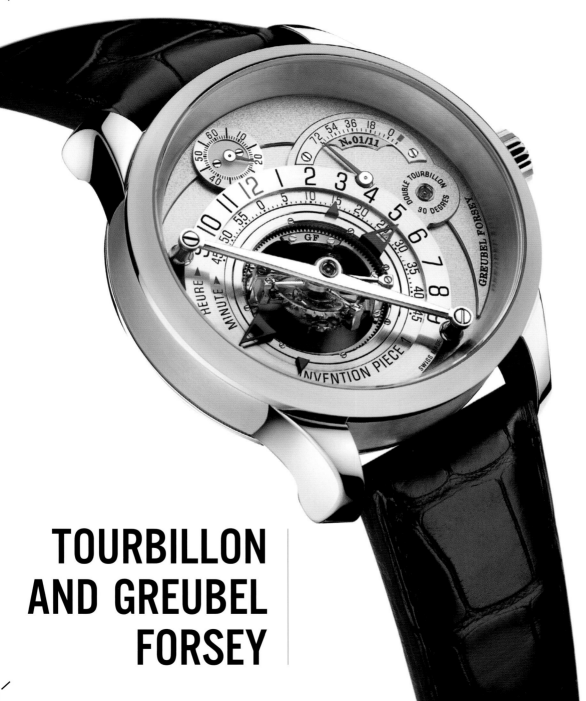

Greubel Forsey Quadruple Tourbillon Secret Calibre GF 03I. Case in platinum with a diameter of 43.5 mm. Leather strap. Indicator for its 50-hour power reserve. Price $830,000. Limited edition of 8 pieces.

TOURBILLON AND GREUBEL FORSEY

A TOURBILLON IS A SMALL COMPLICATED AND COMPOSITE PART IN A MOVEMENT FOR HOUSING THE BALANCE WHEEL AND ESCAPEMENT.

Some time ago, a tourbillon function was considered crucial for the movement's accuracy, in addition to being a symbol of the art of advanced watch making. Modern watches do not require this part, even though luxury timepieces often have a tourbillon to create the impression of being more accurate and exclusive. The finer brands carry tourbillon

The Greubel Fosey Quadruple Tourbillon Secret, in case you should require four tourbillons on your arm.

pieces as their most expensive top models. The idea behind tourbillon was that the effect of gravity would be offset by the mechanics in the watch. Watchmakers believed that a personal watch became inaccurate over time because it was constantly in motion. Countless short stops during the day made the watch become inaccurate. To support this claim, they compared time to the city's many clocks, which were showing the correct time, as supposed to the wristwatches and pocket watches.

It was watchmaker Abraham-Louis Brequet who in 1985 came up with the idea of a tourbillon: a rotating cage to keep the balance in a watch so that gravity would not alter the correct time. The word *tourbillon* means "whirlwind," a direct allusion to the movement. Watchmakers were quick to catch on to this function, and tourbillons became a must for everyone who wanted to make an accurate watch.

Research shows, however, that the tourbillon's effect is miniscule. The fact of the matter is that watches made at the time were poorly constructed and built, but with today's technology and experience, there is hardly any difference between a traditional mechanical watch and a tourbillon. But the precise and detailed machinery in a tourbillon is still a testimony to the watchmaker.

Greubel Forsey is a watchmaker specializing in high-end complicated watches. The company was established in Switzerland in 2004 by Robert Greubel and Stephen Forsey. Greubel Forsey makes movements and watches with several tourbillons and tilted balance wheels for improving precision. Their work was lauded in 2011 when a Greubel Forsey Double Tourbillon Technique won the International Chronometry Competition by Le Locle Museum of Horology. The $500,000 timepiece is for particularly interested people.

Tourbillon in a rectangular case.

Tourbillon.

WATCH ADS

EINSTEIN'S TIMEPIECE WAS A PATEK PHILIPPE

It can be seen in the Patek Philippe historic antique collection. Golden Circles and other classic styles from the current collection can be seen in our latest brochure, on request Suite 629 AD, 10 Rockefeller Plaza, New York, New York 10020

HM4

A traditional wristwatch has a relatively straightforward role: to tell the time. All that is needed is a hand for the hours, another for the minutes, and perhaps a power reserve indicator to keep track of running time. Horological Machine No. 4 Thunderbolt has this—and so much more!

This is not your typical wristwatch. The aviation-inspired case and engine are one. This work of art is incredibly beautiful. HM4 mixes high-tech titanium with a sapphire center section offering a view into the engine.

Thunderbolt is a result of three long years of development. Each of the 300-plus components—including regulator and screws—was developed specifically for this caliber. Horizontally configured dual mainspring barrels drive two vertical gear trains, transferring power to the twin pods indicating hours/minutes and

Watches have a rather simple mission. They show the time and all that is needed is a hand for the hours and another for the minutes. HM4 is not a traditional clock in this respect...

power reserve. The sleek aerodynamic form of the Thunderbolt's envelope has its roots in Maximilian Büsser's childhood passion for assembling model plane kits, though none looked remotely as futuristic as these. The striking transparent sapphire section of the case requires over 185 hours of machining and polishing to transform an opaque solid block of crystal into a complex, exquisitely curved panel allowing light to come in and the beauty of Thunderbolt's engine to stand out. Every component and form has a technical purpose; nothing is superfluous and every line and curve is in poetic harmony. Articulated lugs ensure supreme comfort. Highly legible time is a fringe benefit.

HM4 is available in titanium/sapphire; HM4 Razzle Dazzle and Double Trouble are limited editions of 8 pieces each in titanium/sapphire. HM4 RT is a limited edition of 18 pieces in 5N red gold, titanium, and sapphire.

WATCHES SOLD AT AUCTIONS

JAMES BOND 1973 ROLEX 5513
SOLD FOR $450,000

This watch, made in steel with a steel strap and black bezel, was worn by Roger Moore in the movie *Live and Let Die*. In the movie it had a lot more functions (like a laser cutter and a magnetic field generator). James Bond used it to remove a beautiful lady's dress, and it could even stop bullets.

ROLEX CHRONOGRAPH
SOLD FOR $1.16 MILLION

Only 12 pieces were made of this classic model from 1942. The auction price at Christie's started at $680,000 but ended at $1.16 million. The watch has a steel-colored case with matte silver colored hands and 17 jewels. In addition it has a black tachymeter, a large separate second hand, and a split-second chronograph.

ROLEX GMT 116769TBR
SOLD FOR $485,350

This is one of the most expensive watches ever sold coming from Rolex. It has a diamond-covered bezel with wave pattern and illuminated hands and markers. The strap is made in white 18-karat gold packed with diamonds. Even the case and lock have been decorated with 76 diamonds. This timepiece is automatic and waterproof down to 100 meters.

ERIC CLAPTON'S 1971 ROLEX DAYTONA
SOLD FOR $505,000

In the same way as his guitars, Eric Clapton's watch is highly sought after. This Daytona model has a chronograph function and an unusual silver colored bezel and eyes. It is different from other Daytonas and has been nicknamed "Albino." Eric Clapton's ownership had, of course, a strong influence on the sales price.

PAUL NEWMAN'S ROLEX DAYTONA
SOLD FOR $106,273

This watch has become one of the most classic and sought after Rolex models. The bezel and hands are made in 14-karat champagne colored gold. It has a black bezel with white eyes and a black tachymeter with a white inner ring. This model was made in an edition of only 200 pieces.

ROLEX SUBMARINER FOR CARTIER
SOLD FOR $100,000

An extraordinary tale—Rolex made a few special edition models for Cartier. Cartier sold them in their Fifth Avenue New York store. The first one, a Sea-Dweller, was sold for $91,000 and the second one was this Submariner in gold, which sold for $100,000.

PAUL NEWMAN'S FERRARI-RED ROLEX DAYTONA 6565
SOLD FOR $267,203

This was one of the watches Paul Newman wore in the movie *Winning*. The film is about a race car driver, and Rolex designed the watch with clear references to Ferrari's characteristic red color. This chronograph is 37 mm in diameter and comes in stainless steel with a steel strap.

PATEK PHILIPPE 1928
SOLD FOR $3.6 MILLION

A Patek Philippe watch from 1928 in 18-karat white gold with oval case and single button chronograph was sold for no less than $3.6 million at Christie's in 2011.

FRANK MULLER AETERNITAS MEGA 4
SOLD FOR $2.7 MILLION

This is one of the most complicated watches in the world and has as many as 36 complications and an incredible 1,483 separate parts. The Patek Philippe Calibre 89 was defined as the most complicated watch before Franck Muller took over the title with his Aeternitas.

PATEK PHILIPPE PERPETUAL CALENDAR
SOLD FOR $5.7 MILLION

This classic and beautiful timepiece in 18-karat gold was sold at Christie's for $5.7 million. It has a moon phase, chronograph function, and perpetual calendar. Patek Philippe watches are among the world's most exclusive and sought after by collectors.

JAEGER LECOULTRE AND THE U.S. NAVY SEALS

Developed with marine soldiers and command divers from Navy SEALs.

SOMETHING IS NOT QUITE RIGHT HERE. JAEGER LECOULTRE HAS ALWAYS HAD ITS OWN STYLE IN THE SAME LEAGUE AS PATEK PHILIPPE AND AUDEMARS PIGUET. EXPENSIVE, CLASSIC MECHANICAL MASTERPIECES. A SOBER UNIQUE BRAND FOR EDUCATED WEALTHY MEN IN SUITS. NOT FOR MARINE SOLDIERS AND COMMAND DIVERS FROM NAVY SEALS.

But the fact is that Jaeger LeCoultre teamed up with the U.S. Navy SEALs in developing the Master Compressor Diving Navy SEAL series. It comes in models with alarm and in limited editions.

It is no secret that watch manufacturers are run by their marketing departments and deliberately use celebrities, mountaineers, military, astronauts, and pilots for building image and to influence us consumers. Anyway—this is a great watch in the good Jaeger LeCoultre tradition. The model was introduced in 2011, and was made in close collaboration with American divers and the Navy SEALs, for working in all types of conditions (the acronym SEAL stands for Sea, Air, and Land). This type of equipment must be rugged and reliable. Jaeger LeCoultre started this project by equipping several SEALs with Master Compressor divers' watches and taking note of their feedback. They remarked that the cases and rotating bezels reflected light too strongly and that the surfaces of the watches should be less shiny and more matte. The SEALs also suggested that the construction of the bezels needed rethinking, because they sometimes separated from the cases when the watches were subjected to their tough daily regimes.

Jaeger LeCoultre started using Super-LumiNova, which gives a clear view underwater and in difficult conditions. In the dark the watch glows so brightly and for such a long time that it almost seems as though there is a light source behind it. The bezel clicks cleanly into place in one-minute increments and is easy to use even with diving gloves. The surface of the steel case is entirely matte finished and embellished with a longitudinal, abraded pattern. This gives the watch an attractive technical look and helps reduce glare.

In the watch you will find a Jaeger LeCoultre 899 Caliber ticking with 43-hour power reserve. It shows hour, minute, second, and date, in addition to a diving bezel. The stainless steel case is waterproof down to 30 meters. The sturdy rubber strap can be swapped for a different look. There are several models in the series, with a starting price of $10,000.

Master Compressor U.S. Navy SEALs deserves a good report card. The dial's legibility is especially top notch, and so is its appealing understated look. Thanks to the input from Navy SEALs the watch is well equipped for underwater use, and will not disappoint professional divers or military elite divers anytime soon.

Jaeger LeCoultre Master Compressor Diving Automatic Navy SEALs. Calibre 899 Automatic with 43-hour power reserve, 42 mm diameter, water resistant to 30 bar pressure.

Titanium case, 48.8 mm in diameter, and height 17.9 mm. Rubber protected crown. Sapphire glass and waterproof down to 100 meters. Mechanical winding with its own HYT movement and 65-hour power reserve.

HYT H1

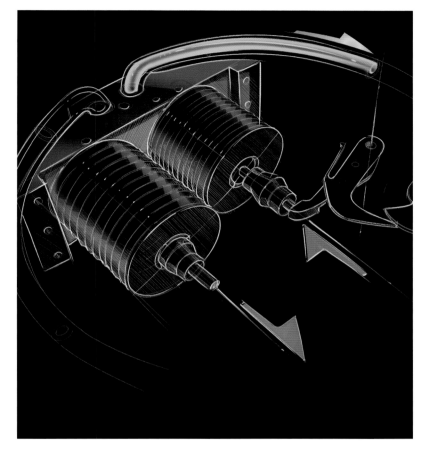

A blend of traditional technology, innovation, and fluorescein-loaded fluid in thin tubes.

HYT H1 IS IN MANY WAYS THE WATCH'S ANSWER TO THE CAR INDUSTRY'S "SUPER CARS," WITH THE SAME STATUS AS THE LAMBORGHINI, FERRARI, AND PORSCHE. LIKE THE SUPER CARS, HANDMADE AND FULL OF TECHNICAL FEATURES AND TECHNOLOGICAL INNOVATIONS, HYT H1 PUSHES THE LIMITS FOR WHAT IS POSSIBLE FOR A MECHANICAL WATCH.

Mechanical watches are small wonders in themselves with hundreds of tiny parts assembled into a small work of art. This watch is taking it even further. It indicates time by the help of moving magnets and fluorescein liquid.

While most parts of the HYT H1 more or less have the same functions as equivalent parts in other mechanical watches, there are some significant differences. Imagine a super car with pistons moving up and down. The similarities couldn't be more striking.

The fluid indicates the hours and the liquid is pumped by hydraulic pressure. The minutes have their own hand, and the watch is surprisingly easy to read. To indicate time correctly and completely, the pipe must be produced with extreme and exact dimensions and have a tolerance of a few nanometers. The reservoir must be filled with a precisely accurate amount of liquid. The tolerance is very minimal, but the performance is impressive. HYT has registered seven patents on the technology and one on the design.

The clock is 48.8 mm in diameter and 17.9 mm in thickness and has a 65-hour power reserve. It displays the time in an elegant way, and is made mostly by hand. Like a super car's, the building process is long, and the HYT H1 is thus produced in a very limited-number series. This is an example of a true high-performance super watch.

THE WATCH THAT UNLOCKS YOUR CAR

Made by designer Jaeger LeCoultre, this exquisite timepiece is known for its light weight and innovative design. The case was made in Grade 5 titanium like the grill on a car, and the digits are covered in rhodium.

The technology in the watch was integrated with the Aston Martin DB9, giving the owner

the option to locate the car and even unlock it by pressing the sapphire crystal control button.

This was the first mechanical watch to be integrated in the operating system of a luxurious car. It is not clear exactly how many pieces were launched, but they're priced at $35,000.

CAR BRANDS AND WATCHES THAT MATCH

BASED ON DIFFERENT BRAND FACTORS SUCH AS
HISTORY, VALUES, STATUS, AND VOLUME AS WELL AS
MY OWN OPINION, I HAVE COMPILED AN OVERVIEW
OF SIMILARITY BETWEEN CARS AND SOME WATCH
BRANDS. THIS IS NOT NECESSARILY THE WATCH
THAT OWNERS WEAR, BUT MORE HOW THE SELECTED
BRANDS MATCH THE DIFFERENT CARS.

Omega	BMW and Audi
IWC	Porsche
Jaeger LeCoultre	Aston Martin
Blancpain	Range Rover
Panerai	Audi RS models
Vacheron Constantin	Maserati
Rolex	Mercedes
Hublot	Ferrari
Patek Philippe	Rolls Royce
Audemars Piguet	Lamborghini
Seiko	Honda
Citizen	Toyota
Sinn	Lotus
Zenith	Saab
Bremont	Jaguar
Victorinox	Peugeot
Invicta	Opel
Breguet	Bugatti
Hamilton	Ford
Tag Heuer	Volkswagen
Grand Seiko	Lexus
Casio	Nissan
Breitling	Lexus
Oris	Subaru

BASELWORLD

Part of the Rolex stand at Basel World 2013. If you want to visit the inside it's by invitation only.

EACH YEAR, THE RENOWNED WORLD WATCH AND JEWELRY SHOW TAKES PLACE IN BASEL, SWITZERLAND. THIS SHOW COMPARES TO THE GREAT CAR SHOWS IN GENEVA AND FRANKFURT, EXCEPT IT IS MAINLY FOR PEOPLE WHO WORK IN THE INDUSTRY.

Even though I am relatively well updated on the different brands and what is happening, I am always surprised to see how many brands I have never heard of before. In 2013 there were more than 1,400 exhibitors from 40 different countries.

This is the venue where all producers meet their distributors and their customers. More than 120,000 people from 100-plus countries visit the show each year. On entering the main hall, you find the most exclusive brands situated there.

Rolex/Tudor has a gigantic building-like stand with a guarded entrance checking your credentials. There is no way you will enter without an invitation or the correct pass.

Tag Heuer, Hublot, Bulgari, Zenith, and Patek Philippe are all superbly situated. Fabulous stands feature offices and press rooms, with crowded restaurants and bars at all times. This is where new deals are made, new models are shown, and the signing of orders takes place. Sometimes it is also where some find out they may not successfully sell a brand. Swatch Group has become more dominant with their brands. Farther along the main hall, you find Longines, Certina, Balmain, Omega, Glashütte, Rado, Tissot, Breitling, and a whole lot more.

The famous brands have large stands, and many of the exhibitors use them year after year.

Everyone fights for attention—gigantic aquariums, astronaut suits, and even a small private plane are exhibited.

Blancpain built a really nice open exhibit honoring 50 years with Blancpain Fifty Fathoms. The entire story was on display, with photos from the first dives, and the diving equipment of the time.

If you move along into the other halls, there are hundreds of small and slightly bigger manufacturers present. It is somewhat easier to have a stand here. These brands work with their customers and at the same time they try finding new distributors in new markets and new countries. I went to visit Rune Bruvik from Bruvik, Norway, to hear how it

plans on selling throughout the market. It is not easy being small and new among the big brands. Some have more than 100 years of history and traditions. I am impressed by the small companies who try to make it. I'm impressed by their enthusiasm, determination, and their large stash of patience for what they believe in.

Baselworld has its own newspaper, and in the press center there are more than 3,500 journalists reporting news, trends, models, and other important stuff in the business. Baselworld's total area is 140,000 square meters, and I know firsthand that it is impossible to cover it all in two hectic days, so be sure to make a list of what you most want to see. This show is time-consuming but a fantastic inspiration!

SMALL SCULPTURES

Dimitriy Khristenko, the U.S.-based Ukrainian-born artist, makes miniature copies of motorcycles and other vehicles (bikes, tricycles, and quad bikes), all built in recycled parts from wristwatches. The strap makes tires while the crystals become windshields or visors.

Khristenko starts by disassembling the watches, then uses all those recycled parts and components in his work to create these fantastic small sculptures.

COMPLICATIONS AND EXPLANATIONS

Complications, in watch terms, means functions in addition to showing hours, minutes, and seconds. A complication may be a moon phase, a chronograph, a month display, etc. Larger complications can be sorted into categories:
— Watches with one or more hands
— Watches with striking mechanisms
— Watches with astronomical indications (calendars)

These three groups are often found in one watch, which results in a large complication.

Each group can also be categorized.

FOR WATCHES WITH ONE OR MORE HANDS
1 — Watches with independent second hands
2 — Watches with jumping seconds
3 — Watches with split-second chronograph
4 — Split-second chronograph

FOR WATCHES WITH STRIKING MECHANISMS
1 — Quarter repeater
2 — Five minute repeater
3 — Half quarter repeater
4 — Small repeater
5 — Full repeater

FOR WATCHES WITH CALENDAR INDICATIONS
— Simple calendars
— Perpetual calendars
— Moon phases

In this book, on page 130, a watch with 36 different complications is discussed. Thirty-six might not sound like a lot, but that makes the Franck Muller watch the one with the most complications.

PERPETUAL CALENDAR
This is a function showing day, month, date, and year. It is automatically adjusted for February and leap years. normally up until the year 2100.

Watches' perpetual calendars are most often based on the Gregorian calendar, so the feature doesn't need correction for more than a century. Other perpetual calendars can be, for example, secular perpetual calendars or Jewish perpetual calendars.

TOURBILLON

This is considered a very special complication in a watch. A tourbillon aims to counter the effects gravity has on the balance of a watch, and thus restores the watch's accuracy.

This complication is usually implemented in expensive timepieces and is exposed on the watch's face to show it off. Modern watches don't really need this, but a tourbillon is simply a novelty and a demonstration of watchmaking virtuosity.

There are more advanced varieties. A gyrotourbillon moves around two axes, both of which rotate once per minute and are run by a special mechanism called remontoire. Triple-axis tourbillon has a third (external) cage with a unique form which provides the possibility of using jewel bearings instead of ball bearings. Quadruple tourbillon is based on two double tourbillons working independently. A spherical differential connects the four rotating carriages, distributing torque between two wheels rotating at different speeds.

HOUR, QUARTER OR MINUTE REPEATER

The minute repeater is a complication in a mechanical watch that tells time by tones chiming when a button is pressed. There are different types of repeaters, from the ones giving hours and quarters, to the ones that by using separate tones tell the time in minutes also. Before free access to electricity, this helped with telling the time in the dark; it is now also helping the visually impaired.

POWER RESERVE

Mechanical watches are normally wound up mechanically or automatically. For a mechanical watch to work, it must be at least 30 percent wound up and be worn around 10 to 15 hours before it is fully charged. The power reserve function shows how long until it will need a new winding.

ALARM

A watch with an alarm function can be mechanical or quartz. There are many inexpensive quartz watches with alarm function. A mechanical alarm beats with a small hammer and is more complicated than traditional quartz alarms.

The adjustable alarm is the traveler's constant companion for wake-up calls, or a valuable tool for the modern businessman reminding him of a meeting, arrangement, or even when his parking meter has expired.

CHRONOGRAPH

There are many watches with this function today. You can easily spot a chronograph by the three small extra hands on the dial and the two push-buttons on the side of the case. A watch with a chronograph is able to measure independent time intervals. A simplified way of describing it? A stopwatch inside your watch. This is shown by small hands on the dial. Normally there are three small hands for seconds, minutes and hours.

RATTRAPANTE – SPLIT-SECOND CHRONOGRAPH – DOUBLE CHRONOGRAPH

This is the most advanced chronograph. Instead of the ordinary stopwatch with seconds, this has two hands. One is on top of the other. When the chronograph is activated, both hands start, but by pressing the split-second button, one will stop while the other continues, to give you the option of timing two things at the same time. By pressing the button again, the stopped timer will automatically resume and they continue together. This is a complex mechanism and it requires the watchmaker to know what he is doing.

TACHYMETER

Located on the bezel, this scale is used to calculate speed based on travel time or to measure distance based on speed. For example, you start the chronograph when an object being measured is passing a starting line, and then stop it when the object reaches the next mile or kilometer marker; the point indicated on the tachymeter scale gives the speed (in mph or kph) of the object.

GMT

Greenwich Mean Time (GMT) is the basis of every world time zone. In watchmaking, the term is used for a 24-hour watch, one that is particularly useful for aviators, astronauts, and the military, or anyone who uses a 24-hour clock. GMT is also commonly referred to as Coordinated Universal Time (UTC) and Zulu Time. A 24-hour watch with an extra hour hand can show two time zones.

MOON PHASE

Moon phase is a complication that used to be quite popular. It is often shown through a screen on the bezel. It shows the current moon phase. This is a complication that can also be found on relatively inexpensive watches.

FLYBACK

A flyback chronograph is a chronograph with a little twist. Normally you need to stop a chronograph to reset it. The flyback function allows you to instantly restart the chronograph with a push of the button at the 4 o'clock position. The hand returns immediately to zero and starts again. (Without this function, the operation requires three separate movements.) The flyback function is closely linked to functions used by pilots and in aerobatics, which require fast and precise timing of maneuvers.

DUAL TIME OR MULTI TIME

A watch with the capability of displaying time in two different time zones at the same time. This can be done either with a small watch inside the watch (in the bezel), or via a watch you can turn or rotate. Jaeger LeCoultre's Reverso is an example of this.

EQUATION OF TIME

This is not an especially useful function nor a very common one. The actual length of a day, in relation to the sun, is not exactly 24 hours; the solar time can be ahead as much as 16 minutes or behind as much as 14 minutes. This rather unusual, and expensive, complication shows the difference between nature's time—the earth's elliptical track around the sun—and human time.

JUMPING HOURS

This is a more than 100-year-old design that is not used much anymore. While ordinary watches have two or three hands moving, this design has a wheel rotating and showing the numbers from a small disc, very often located in a small window or on a screen on the bezel. There are still some retro-inspired watches with jumping hours. Several of the brands mentioned in this book have had models with this technical function.

REGATTA AND DIVING WATCHES

Regatta watches have a countdown function with alarms at the sixth, fifth, and last minute before a race's start. Some more advanced ones show the tide times as well, even though not totally accurately. A lot of people use watches for diving, and these are traditionally large, with rotating bezels for measuring the time of the dive. In addition the manufacturers have developed advanced watches with functions that indicate decompression and stops to decompress. Some watches are equipped with an alarm showing you when to ascend. A diving watch should be waterproof down to a minimum of 200 meters.

"Time is what we want most,
but what we use worst."

William Penn

PHOTO CREDITS AND COPYRIGHT

I would especially like to thank Marine Lemonnier Brennan at Hublot, Emilie Jacquot at Parmigiani, Nicole Banning at IWC, Daniel Boun and Sasi Langford at Project X, Zenith, Swiss Time, Rune Bruvik, Jake Ehrlich at *Jake's Rolex World Magazine* and Tom Møller Christensen at Watchmaker Thorbjørnsen for all their support, help, and cooperation.

I have contacted almost all the different manufacturers to gain access to images, check references, and fact check text. Some have been incredibly helpful in assisting me and I would like to thank them for all their help. Others have a policy to not contribute to various book projects, films, or articles, while others failed to answer.

For some of the images used in this book we have not been able to locate an owner or licensee. If, in this context, I have violated any copyright it has been involuntary and unintentional. If you can supply any information about copyright holders who are inadvertently omitted here, please tell me by contacting my publisher.

THANKS

Thanks to an enthusiastic bunch from Aldente, and Bjørn and Kjell from Jæren Publishing. Thanks to Geddon for the loan of the smashing suit and Alex Hart for the tie and scarf from Made with Hart. To Anne Espeland for feedback during the process, and all the work with translating the book to English. To photographer Fredrik Ringe for the fabulous cover photo and to my good friend, Christian Storm Trosdal, for being a supermodel. The result: 100%.

Hope you agree.
—Ivar Hauge Line